FOOD WRAP

berry
SPARKLING
ICE
BERRY FLAVORED
12 FL OZ (355 ml)
Contains 5% Fruit Juice
apple
SPARKLING
ICE
APPLE FLAVORED
12 FL OZ (355 ml)
Contains 5% Fruit
cherry
SPARKLING
ICE
12 FL OZ (355 ml)
Contains
peach
SPARKLING
ICE
PEACH FLAVORED
12 FL OZ (355 ml)

FOOD WRAP

PACKAGES THAT SELL

STEVEN HELLER & ANNE FINK

Graphic Details

AN IMPRINT OF

PBC INTERNATIONAL, INC.

Distributor to the book trade in the United States and Canada
Rizzoli International Publications
through St. Martin's Press
300 Park Avenue South
New York, NY 10010

Distributor to the art trade in the United States and Canada
PBC International, Inc.
One School Street
Glen Cove, NY 11542

Distributor throughout the rest of the world
Hearst Books International
1350 Avenue of the Americas
New York, NY 10019

Library of Congress Cataloging–in–Publication Data
Heller, Steven
Food Wrap: packages that sell / Steven Heller, Anne Fink
p. cm.
Includes index.
ISBN 0–86636–394–7 (pbk ISBN 0-86636-533-8)
1. Food—Packaging—Design. 2. Labels—Design.
I. Fink, Anne II. Title
NC00.P33H46 1996 96–3545
741.6—dc20 CIP

CAVEAT– Information in this text is believed accurate, and will pose no problem for the student or casual reader. However, the author was often constrained by information contained in signed release forms, information that could have been in error or not included at all. Any misinformation (or lack of information) is the result of failure in these attestations. The author has done whatever is possible to insure accuracy.

Designed by Kristen Lilley

Color separation, printing and binding by Dai Nippon Group

10 9 8 7 6 5 4 3 2 1

Printed in Hong Kong

To Nick,

with the ravenous appettite.

Contents

Consorzio
Rosemary
olive
oil
Net Wt. 375 ml (12.6 Fl oz)

OK.
OK.

TOPAZ
TOFFEE

NOV 11
OPEN
VERMONT
FAMILY FARMS
FARMER'S CHOICE
GRADE A
MILK
Pasteurized • Homogenized • No Additives
HALF GALLON (1.89L)

Introduction

Food packages are the weapons in the war to win a consumer's loyalty. With the huge investment required to develop new and compete with older brands, strategies are employed that would rival many military operations. For every victorious product scores of others are defeated. For every popular brand, armies of tacticians, designers, and marketing experts expend huge amounts of energy devising the perfect package.

A supermarket is stocked with so many competitors that a shopper must be conditioned to buy a specific brand. This special allegiance, or brand loyalty, is hard won. Only after a constant barrage of sales pitches will a shopper even so much as think to try something new. So a three-pronged attack is the most common strategy: advertising, promotion, and packaging. The first introduces and positions the product; the second insures that the brand name becomes a household word, and the third seduces the shopper. In this equation the package is not simply a protective container, it is a signpost—indeed a billboard. Although sustained interest in a brand inevitably relies on quality (a bad product may win a battle but invariably lose the war), without the benefit of these three combined forces a product is as good as invisible. What this strategy must insure is engagement with the eye and the senses through appetite appeal.

Everyone has experienced the seductive sway of an hypnotic aroma. The unmistakable smells of fresh brewed coffee, sizzling bacon, and grilled hash browns make the mouth water so uncontrollably that restaurants have been known to set exterior exhaust fans to high, filling the street with breakfast emanations that lure in a hungry customer. Likewise, a food package must appeal to that part of the brain where appetites reside; where one is most susceptible to suggestion. The challenge of the contemporary package designer is, therefore, how to make paper, tin, cardboard, and glass into something seductive. Transforming unappetizing materials into a banquet is not easy.

From the introduction of the modern commercial package almost a century ago a variety of conventions remain in place. The most common is a photograph or painting of a prepared packaged food. A stack of pancakes soaked in butter and syrup, a steaming bowl of soup, and plate of spaghetti topped with a delectable sauce excites the salivary glands; reproductions of fruits and vegetables, a freshly sliced apple, orange, or pear on a container of juice, or dew soaked berries on a jar of jelly or jam, have incalculable subliminal powers. In the absence of real aromas precisionist images like these are mouth watering substitutes. The fact is, without these facsimiles the supermarket would be little more than a warehouse of generic staples. With them, it is a cafeteria of tasty beckoning displays.

Food packages do not, however, show cooked foods or nature's bounty exclusively. The contemporary package is designed to infiltrate the consumer's psyche on a variety of social levels, appealing to the shopper's lifestyle, as well as societal and health concerns. These days a product cannot just be mouth watering or thirst quenching but necessary for the body, mind, and soul. And so old packaging conventions are pushed, taboos are shaken, and standards dislodged to redefine the relationship between consumer and product. In the 1950s virtually all supermarket brands, regardless of the product, looked basically the same: bold gothic type, loud primary colors, and friendly (often goofy) trade characters and mascots on a label or box. They were designed by specialists who understood marketing conventions and taboos. Today, a wide range of graphic

designers (many who do not consider packaging to be their specialty) apply themselves to packages as one part of a general practice. Yet because of these designers, the old school conventions have been challenged and taboos busted. In the eighties bold gothics gave way to both classical and fashionable typefaces; loud primary colors were replaced by subtler pastels; and trade characters were streamlined or diminished in size. Yet despite changes in graphic style, in the final analysis food packaging is still the least experimental and probably the most conservative of any graphic design form.

Consumers insist that they want novelty as a respite from daily routines, yet an analysis of the average shopping cart reveals that few purchases are novelty or impulse buys. Consistency is the watchword, if not the hobgoblin, of food industry graphics. Once a shopper is hooked and loyal to a particular brand the emphasis of design is on maintaining their loyalty—not confusing them with a barrage of new graphic motifs. Even when "new and improved" versions of old brands are released the product's fundamental graphic schemes are continued. Venerable brands may have periodic makeovers but few manufacturers are foolhardy enough to threaten their hard-earned equity by tampering with success. Only those older products facing a diminishing consumer base engage in radical cosmetic surgery, and only after extensive market research determines that it is feasible and necessary.

Consumers really want their brands to be loyal to them. They don't want surprises that interrupt their basic routines or, more importantly, strain their budgets. Yet this should not mitigate the natural tendency to experiment with new things. Although staple purchases usually remain constant, fashionable or fun foods—a growing segment of the current market —are where new brands with quirkily designed packages have a chance of winning a share. In recent years salsas, bottled waters, fruity soft drinks, and "natural foods" have invaded and staked out territory, first in the specialty food shops, and next in many larger supermarkets. This is where shifts in package trends begin to occur. If smaller, eclectic brands are successful in capturing consumers larger ones may be influenced, too. This is also where contemporaneity is a virtue; the more fashionable the package the more likely it is to attract the impulse buyer. Like turn-of-the-century posters, which were aggressively collected in their day, the more stylish the package, label, or bottle, the better chance it will have to be displayed with pride, not just hidden away in the kitchen cabinet.

Food packaging representing a lifestyle is more than a protective container, it is a status symbol every bit as charged as a Mercedes emblem or as commonly worn as the Nike swoosh. In the eighties the intense competition in the food industry has forced the repositioning of various foods as more than just consumables—they are objects, too.

Advertisements for average bottled waters claim to quench thirst, but packages for the higher priced premium brands presume to satisfy social impulses. To impress one's guests, the theory goes, seltzer is too plebeian, so spend more for a bottle that will stand alongside a decanter of the choicest Beaujolais Nouveau. With its elegant cobalt blue bottle Ty Nant water, for example, is perhaps even more stylish than the most fashionable wine. And the price is accordingly high. In the eighties specialty foods, once the province of small, exclusive caterers and food boutiques, have found their way into the supermarket where shoppers are able to choose between the quotidian and premium. Sometimes the only difference between these two genres is the look of the packages, and so design has become a tool in positioning the brand. In recent years the trend in high- and low-end packaging has influenced the overall look of contemporary package design.

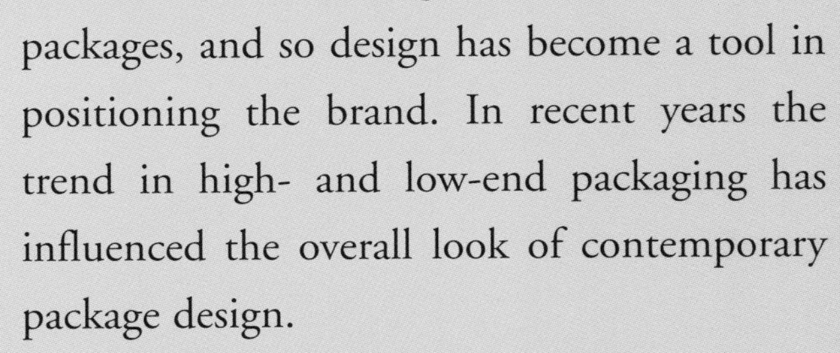

Package design has evolved from the terracotta vessels of antiquity to the recycled paperboard of today; the graphics have come a long way since originating three thousand years ago when the earliest known papyrus labels were used to identify contents. But packages are still fundamentally rooted in the 15th century when paper could be produced inexpensively and product marking begins. While the form varies the function is ostensibly the same. The history of the package provides fascinating contrasts and similarities with current practice.

From the 1500s to the late 1700s printed labels made from handmade paper and printed on wooden presses were primarily used for bales of cloth and drug vials. Production changed in 1798 when the Frenchman Nicolas-Louis Roberts invented a paper making machine and the Bavarian Alois Senefelder developed the lithographic process. Labels soon became a common feature on packages. They were colored by hand until 1835 when the Englishman George Baxter discovered a process to print multiple colors from wood engravings onto a monochrome base. He patented this method and by 1850 his process of chromolithography made it possible to print twelve different colors on a single label.

There was an explosion of printed items throughout America after the Revolution. Every pea and oyster box was branded with a company identity. This kind of packaging flourished for over a century, although it was not until an American, Stanton Avery, created and produced the self adhesive label in 1935 that a key development in packaging history begins. Labeling became a huge industry, and anything that was not printed on a box was given a label. These paper labels, with pictures of produce, trade characters, or happy consumers, tended to stain and peel and are now replaced by ones which repel oil, water and grease.

In the 19th century, illustrations of distant lands and exotic people lured buyers into believing that a common product was imbued with mystery, indeed romance and adventure; their purchase was not commonplace, but precious and rare. In addition to the exotic graphic conceit, testimonials from ersatz experts attested to a product's beneficial qualities. Though these popular sales ploys were usually bogus, what consumer could ignore the wisdom of a doctor or professor, even if these experts earned their degrees from advertisements in pulp magazines. The exact method may be different, but many mainstream contemporary food packages continue the tradition of paid celebrity testimonials.

With the advent of mass consumerism in the late nineteenth century, packages were transformed from quaint decorative objects into no-nonsense tools of persuasion. Nor were they merely commercial vessels but containers of truth — indeed discovery. "For manufacturers, packaging is one of the crucial ways people find the confidence to buy," writes Thomas Hine in *The Total Package, the Evolution and Secret Meanings of Boxes, Bottles Cans, and Tubes* (Little Brown & Co., 1985). "It can also give a powerful image to products and commodities that are in themselves characterless." The package, label, and bottle were at once the totems, icons, and reliquaries of great essences. When individual packages replaced huge bags and bushels that once filled the old general store, the physical package itself symbolized progress. Today, graphics tell the story.

Package design does not exist in a vacuum but is one component of a system that controls consumer behavior. In most cases, the shopper has already been influenced by print advertisements, thirty-second television commercials, radio jingles, and coupon and premium promotions. "Advertising leads consumers into temptation. Packaging is the temptation," continues Hine. What makes a tempting package? In the course of researching Food Wrap we decided that the majority of mainstream products no longer have this particular quality, although presumably they once did. Most venerable packages have become little more than visual noises that stimulate Pavlovian responses. A box of Kellogg's Special K, for example, no longer has the same surprising allure as it did when first issued decades ago, but it still engenders immediate recognition. The same is true for scores of other well known brands whose packages are as effective as traffic lights in telling the consumer to stop and go. But are they really tempting? Or are they so endemic to the shopping ritual that the consumer selects them without thinking?

In a marketplace overstocked with familiar packages, only the novel, unique, or out-of-the-ordinary offer any real excitement, which is what governed the selection of the packages shown in *Food Wrap.* On the following pages are designs created (or at least introduced) within the past five to ten years, a period of growth in the specialty food industry, which have challenged the common truths about package design. While they are all quite practical, there is an overarching sense that these are appealing to a new breed of consumer, and an adventuresome older one. The quality of the design, matters of type, illustration, style, etc., not the comparative success of the product in the marketplace, determined whether the editors included the specimen or not. With design as the qualitative standard, the organizing principle is based on how visual language is employed. Within the past decade we have found that food packaging

neatly falls into six relatively thematic categories, which we call, "Oh Natural" (packages that either focus on the natural benefits of the product, or use natural or recycled materials); "Home Spun Values" (the return to or simulation of the idea of products made in the home, or in close proximity); "Nostalgia and Vernacular" (similar to the home spun, but more specifically trading on fashionable or kitsch, once passé design styles); "Joke and Jest" (humor as a presentation tool); "Classical Dress" (elegance based on the traditional tenets of typography); and "Post Modern Dress" (the most contemporary, convention busting applications of graphic design).

The packages in *Food Wrap* represent two objectives: Those establishing a permanent position in their segment of the market, and those trading on particular fashions or trends for momentary impact. We have tried to mix relatively rare and arcane packages together with some very well known, but exemplary, national brands in an attempt to show how the eclecticism of today's design sensibility is having widespread influence, and how some of the more "radical" approaches of the specialty items have rubbed off on the mainstream. *Food Wrap* is, moreover, an admittedly subjective chronicle of packages that transcend the commonplace. Although some of these products may lose the war to win brand loyalty, we publish them here so that despite how they fare in the market designers can appreciate the packages that possess real appetite appeal.

Post Modern Dress

Grunge, Hip Hop, Rave—alternative graphics born of the current music, art, and cultural scenes—have found their way off the street and onto supermarket and specialty food store shelves. Packages using contemporary graphics, including edgy type, layered images, and discordant color combinations, are targeted at teenage and twenty-something customers. Under the broad rubric of Post Modern, emblematic foods and drinks are being aggressively marketed to the next wave of dedicated consumers.

Like the youth culture magazines and posters that revel in this new visual language, food products with Post Modern veneers seem to assert their individuality. But in fact, such packaging is designed to appeal to an emerging mass that has accepted a new vernacular which has already been market tested to the hilt before reaching the stores. Food packaging is rarely the wellspring of innovation, but in recent years various innovative graphic design approaches have had wide exposure through such packaging. A case in point is the Chaos brand of ice tea introduced in the early 1990s during a surge of quirky new soft drinks, notably fruity and herbal flavored ice teas. A melange of circles, ellipses and triangles orbiting around futuristic typography and set against fluorescent colors gives Chaos a techno image that suggests rebellion, which bears little relationship to the beverage inside. Nevertheless, the imposing, extra large can is not merely a container, it is a virtual fashion accessory. Indeed the goal of Post Modern packaging is not only to attract the consumer's attention but infiltrate his or her lifestyle and ultimately demand brand loyalty, if only for a brief generational moment. Similarly, the Fruitopia brand of teas and fruit combos, bottled by Coca-Cola (the most mainstream beverage on the market), appropriates grunge-inspired primitive graphics influenced by *l'art brut*. Using metallic papers and fluorescent colors, the brand's identifiable kaleidoscope of abstract images which challenge conventional beverage graphics have become a new standard in new wave commercial iconography. Much in the same manner, Groove Dots, a hip new chocolate, is packaged for the fashion conscious consumer. White typography and circles float above blue and orange rectangles on the box. The contrasting colors collide with the white dots, complementing the products' playful name.

Not all Post Modern graphics are grunge or rave-inspired. The smorgasbord of contemporary style that designers currently sample includes various sub-genres which quote neo-classicism, punk, retro in graphic mannerisms that are a few parts old, a few parts new, and when mixed together, totally contemporary. Tra Vigne olive oil and Chateau Nicholas maple syrup are dressed in more subtle Post Modern shapes and hues, but are every bit as con-

tempo as more radical packages. The graphics are not the only distinguishing characteristic, the elongated bottles for Tra Vigne and Chateau Nicholas exude a fashionable, indeed attention grabbing, aura. These new brands literally rise above the more familiar ones.

Another Post Modern method to attract customers to new brands is by agressively overturning marketing conventions by placing quotidian products in unfamiliar packages. Coffee is usually sold in cylindrical aluminum cans with plastic lids, but the tony Cafe Society coffee is packaged in white squishy bags reminiscent of potato chip packages. In the supermarket the shopper may ask if this product has been placed in the wrong aisle, but on second glance the colorful palate and dancing typography has stolen the show.

Stealing attention is the name of this packaging game. That is why in the '90s many new packages are eschewing the traditional, orderly grid structures in favor of more cacophonous graphic styles. Typography is quirkier, sometimes layered, and on rare occasions almost illegible. Expression has become the watchword, and complexity the conceit. The products in this section embody the controlled chaos of the commercial Post Modern sensibility.

Consorzio Flavored Oil

Client
Napa Valley Kitchens

Designer/Illustrator
Michael Mabry

Photographer
Michael Lamotte

Cafe Society Coffee

Client
Cafe Society
Coffee Company

Firms
Cafe Society
Coffee Company
C[2]

Art Director
Laurie J. Sandefer

Designer
Carrie Clammer

Photographer
Dirk Albarez

Gourmet Snackin' Sauce

Client
Amazing Grazing

Firm
Haley Johnson Design Co.

Designer/Illustrator
Haley Johnson

Photographer
Paul Irmiter,
Irmiter Photography

Italia Packaging

Client
Italia Restaurant

Firm
Hornall Anderson
Design Works

Art Director
Jack Anderson

Designers
Jack Anderson
Julia LaPine

Illustrator
Julia LaPine

Typographer
Thomas & Kennedy

Photographer
Tom McMackin

ITALIA
ITALIA

ITALIA
ITALIA
ITALIA

Italia Collateral Materials

Client
Italia Restaurant

Firm
Hornall Anderson Design Works

Art Director
Jack Anderson

Designers
Jack Anderson
Julia LaPine

Illustrator
Julia LaPine

Typographer
Thomas & Kennedy

Photographer
Tom McMackin

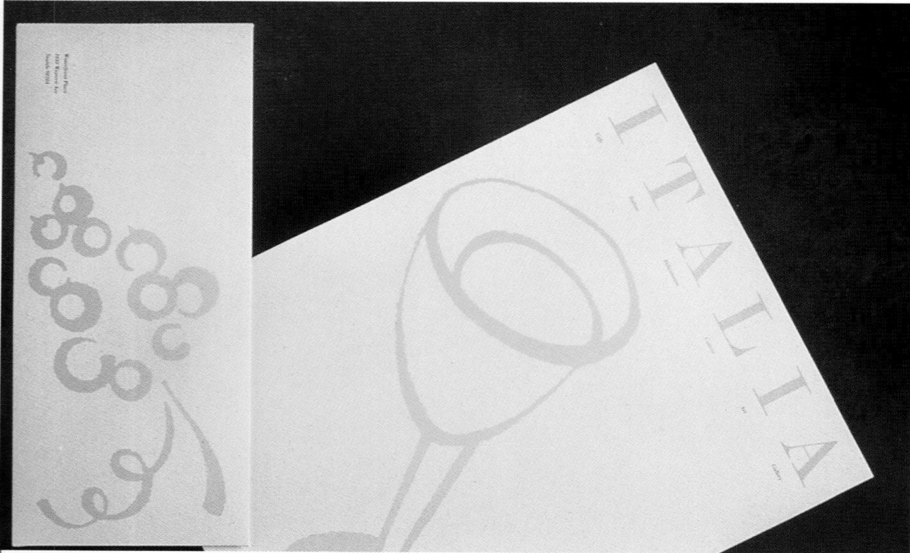

Allegro Tea

Client
Allegro Coffee Company

Firm
Vermilion Design

Art Director
Robert Morehouse

Designers
Leslie Blanton
Brad Clemmons

Photographer
Brian Mark

Allegro TEA
Estate Ceylon (Sri Lanka)
FULL LEAF LOOSE TEA
This superb high-grown Ceylon has a captivating honeysuckle aroma and sweet, penetrating flavor.
NET WT. 4 OZ. 112g
BREWS SIXTY CUPS
TEAS • THE FINEST FULL LEAF LOOSE
Allegro TEA
ALLEGRO TEA
Allegro TEA
Silver Tip Jasmine Yin Hao
FULL LEAF LOOSE TEA
One of the finest scented teas in the world with an ethereal aroma and captivating flavor.
NET WT. 4 OZ. 112g
BREWS SIXTY CUPS
Allegro TEA
China Gunpowder Temple of Heaven
FULL LEAF LOOSE TEA

NET WT. 12 OZ. 340 g
FRESH ROASTED
WHOLE BEAN
Allegro COFFEE
ALLEGRO BLENDS
Allegro Blend
An upbeat blend of Indonesian and Central American coffees. Very smooth and full-bodied with a lingering, spicy flavor.
NET WT. 5 LBS.
Allegro COFFEE
ALLEGRO BLENDS
Allegro Blend
Allegro COFFEE
ALLEGRO SINGLE ORIGINS
Kenya AA
SELECTED ESTATES
Allegro COFFEE
CERTIFIED ORGANICALLY GROWN
Organic French Roast
A dark-roasted blend, roasted to bring the natural oil in the beans to the surface.
Allegro COFFEE
VARIETY
Allegro FINE COFFEES
AZTEC HARVESTS
GOOD COFFEE FOR A GOOD CAUSE
GROWN WITHOUT CHEMICALS
ORGANIC MEXICAN ALTURA COFFEE
DECAF
Café Amico
100% ORGANICALLY GROWN DECAF
Grown at 5,000 feet elevation, these OCIA certified organic beans brew a balanced, mild cup.
Net wt. 12 oz.
Good Coffee for a Good Cause
Allegro FINE COFFEES

Allegro Coffee
Packaging/Identity

Client
Allegro Coffee Company

Firm
Vermilion Design

Art Director
Robert Morehouse

Designers
Leslie Blanton
Brad Clemmons

Photographer
Brian Mark

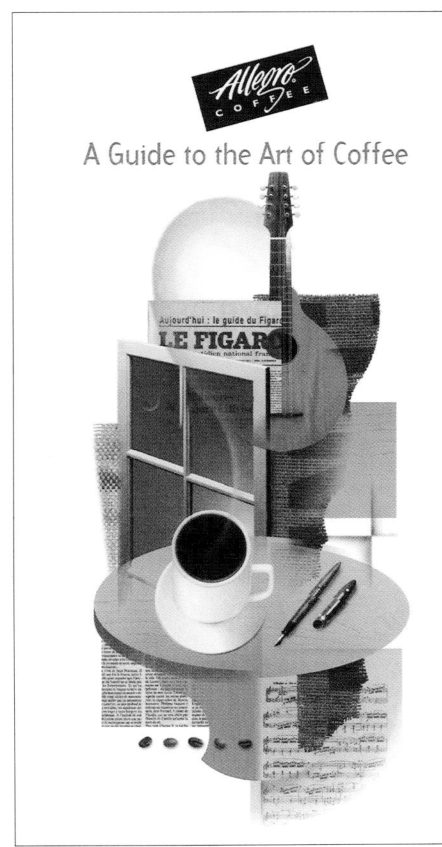

Perrier

Client
The Perrier Group
of America

Firms
Lipson–Alport–Glass
& Associates
Studio Bustamante

Art Director
Elliot Schreiber

Designer
Lori Harast

Illustrator
Gerald Bustamante

"The Art of Refreshment"

Client
The Perrier Group
of America

Firm
Lipson-Alport-Glass
& Associates

Art Director
Sam J. Ciulla

Designers
Carol Davis
Tracy Bacilek
Amy Russell

Illustrators
Terry Allen
John Jinks
Jonathon Lund
David Diaz
Teresa Cox
Tatjana Krizmanic
Ann Field
Coco Matsuda
John Nelson

GROOVE DOTS
AND HOW!
YOWZA!
groove dots
GIMME SOME!
GROOVE DOTS

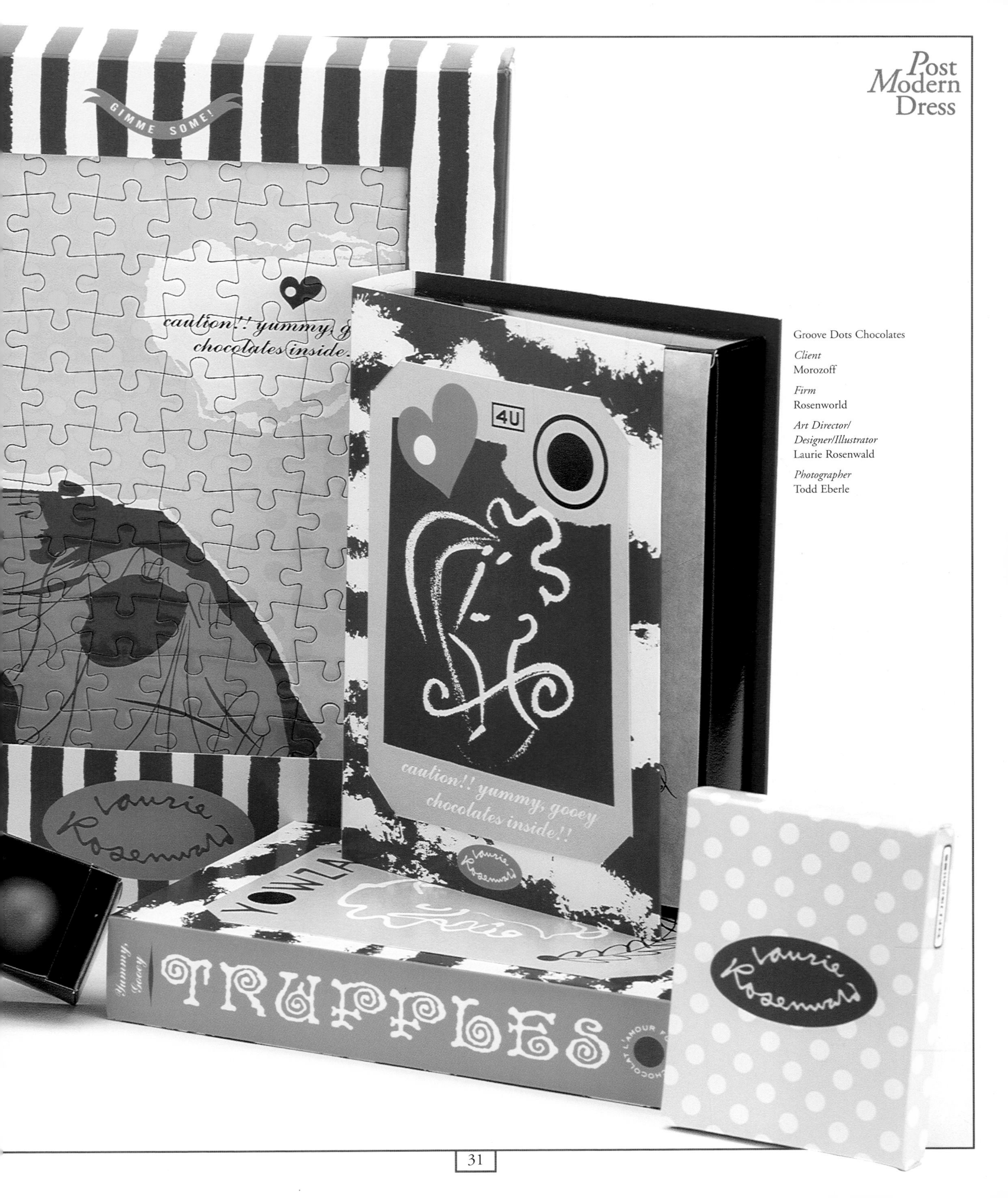

Groove Dots Chocolates

Client
Morozoff

Firm
Rosenworld

Art Director/ Designer/Illustrator
Laurie Rosenwald

Photographer
Todd Eberle

Snelgrove Yogurt
and Ice Cream

Client
Snelgrove

Firm
Hornall Anderson
Design Works

Art Director/Illustrator
Julia LaPine

Designers
Julia LaPine
Jill Bustamante

Photographer
Tom McMackin

Dulce Pecan Pralines

Client
El Paso Chile Company

Firm
Laster & Miller

Art Director
Nancy Laster

Designer/Illustrator
Julian Rivera

Travigne

Client
Real Restaurants

Firm
Michael Mabry Design Inc.

Art Director/ Designer/Illustrator
Michael Mabry

Photographer
Michael Lamotte

Chaos

Client
The Stroh
Brewing Company

Firm
Duffy Design

*Art Director/
Designer/Illustrator*
Neil Powell

natural
NO PRESERVATIVES
Drink this.
Chaos
iced tea
RASPBERRY FLAVOR
rad berry
24 fl. ounces
(1PT. - 8 FL. OZ.)

natural
NO PRESERVATIVES
Drink this.
Chaos
lemonade
LEMON FLAVOR
edgy lemon
(it's a lot to swallow)
24 fl. ounces
(1PT. - 8 FL. OZ.)

natural
NO PRESERVATIVES
Chaos
lemonade
STRAWBERRY FLAVOR
straw 'mon
it's a jungle out there.

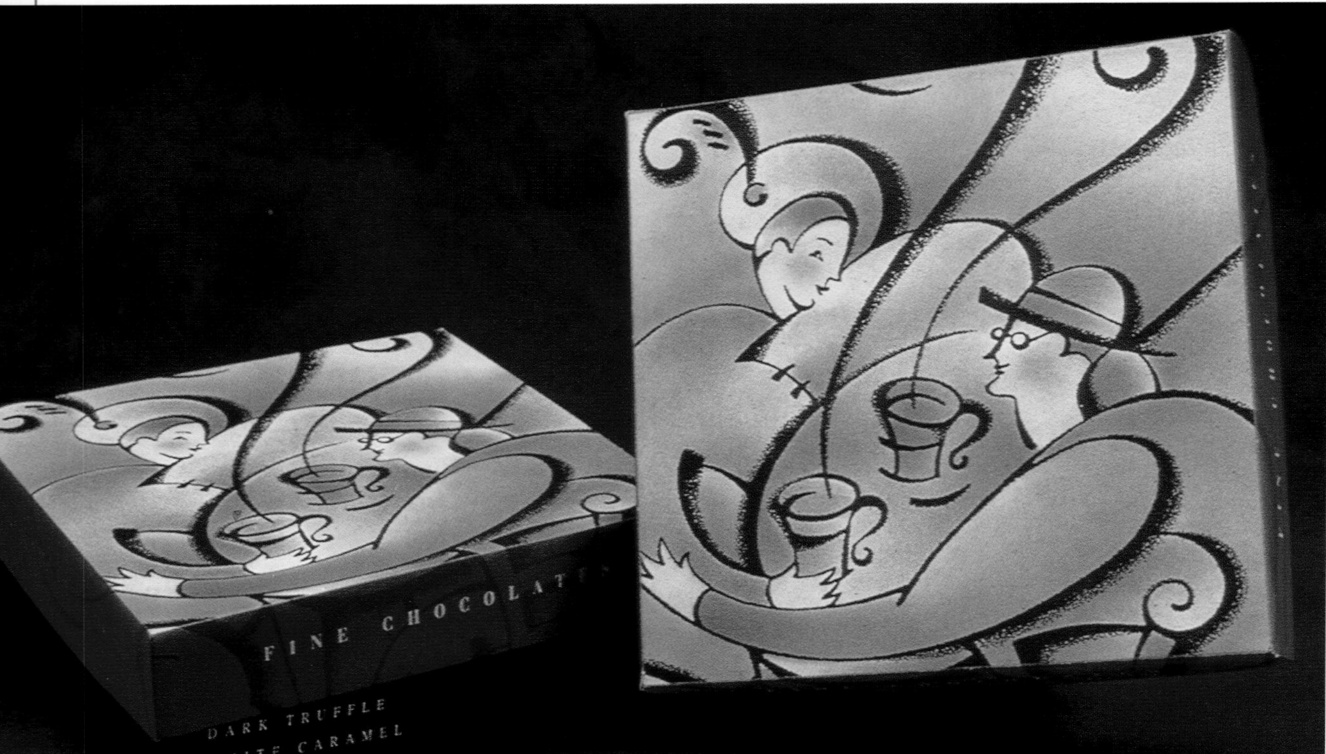

Starbucks
Home Espresso Kit,
Gingerbread Kit,
Fine Chocolates,
Toffee, and Collateral

Client
Starbucks Coffee Company

Firm
Hornall Anderson Design Works

Art Director
Jack Anderson

Designers
Jack Anderson
Mary Hermes
Julie Lock
David Bates

Illustrator
Julia LaPine

Photographer
Tom McMackin

STARBUCKS COFFEE
STARBUCKS COFFEE
STARBUCKS COFFEE
STARBUCKS COFFEE
STARBUCKS COFFEE
STARBUCKS COFFEE
STARBUCKS COFFEE

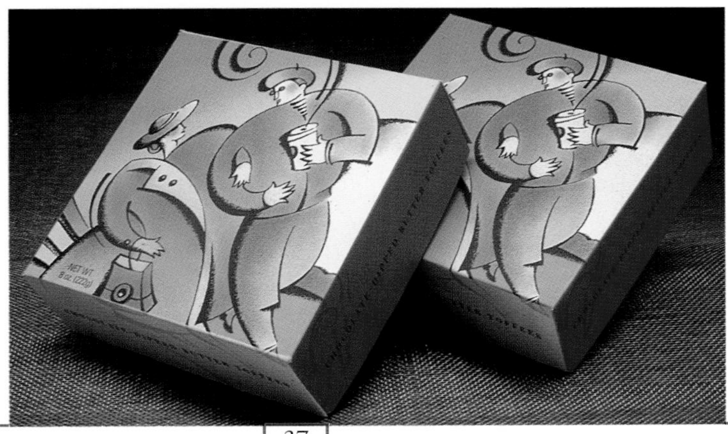

Cloud Nine

Client
Cloud Nine, Inc.

Firm
Haley Johnson
Design Co.

Designer/Illustrator
Haley Johnson

Photographer
Paul Irmiter,
Irmiter Photography

Vermont Gold
Gift Box

Client
Chateau Nicholas

Firms
Zu Design
Chateau Nicholas

Art Director
Peter Vogel

Designer
Taro Masuda

Illustrator
Anthony Russo

Photographer
Chris Vaccaro

Racheli's Identity

Client
Racheli's

Firm
Grafik
Communications Ltd.

Creative Director
Judy Kirpich

Designers
Michael Shea
Susan English

Photographer
David Sharpe

Taru Table Tent

Client
Heublein

Firm
Halleck Design
Group

Designers
Ross Halleck
Robt Kastigar

Joke *and* Jest

"Make 'em laugh" sang Donald O'Connor in Singing in the Rain. *Although he was referring to a theater audience, the sentiments might extend to the supermarket shopper as well. Consciously or not, consumers want to be entertained while conducting their mundane chores in the marketplace. The most entertaining packages have the potential to grab the shopper's attention, and when it comes to making food packages entertaining, the most effective way is through wit and humor. But it's also the most difficult. In the crowded marketplace brand loyalty is often an impenetrable barrier between new and old products, and in this struggle to capture consumer interest wit is a weapon that could easily backfire.*

Graphic wit is hard enough to achieve when the constraints imposed by such a difficult marketplace do not exist in force. Generally with graphic design the fine line between true wit and crude slapstick is easy to cross. Silly sight gags and tired visual puns are rampant in design, which is why intelligent and original graphic humor has such a strong and memorable impact. When it comes to food packages the ability to inject joke and jest is even harder. Food is not a laughing matter, and although the shopper wants to be entertained, their first order of business is to identify the most appropriate, functional, and economical products. A brand of, say, barbecue sauce or lemonade will usually not sell (more than once, that is) based on witty packaging alone, however, a humorous logo, illustration, or decoration may very well encourage the consumer to consider it. In such a highly charged, competitive environment even a second is considered a triumph.

But if the package makes the consumer roll over in the supermarket aisles, humor will not necessarily convince him or her of product value. Humor is not always the correct strategy, so when it is used it mustn't be so far-out so as to obscure its function or appetite appeal. Perhaps the most effective use of wit in packaging are for impulse-buys, such as snacks, ice creams, soft drinks, and other so-called fun-foods. Humor has also proved appropriate for food companies attempting to establish personalities through varied product lines.

Park Kerr of the El Paso Chili Company is an expert in the art of graphic jest. His flourishing and expanding line of "fancy foods" use humor as a key element in the identity of popular, trendy items, such as salsa, tortilla chips, and spicy relishes. He draws from a stable of leading designers, Michael Mabry, Charles Spencer Anderson, Seymour Chwast, Louise Fili and Jenifer Morla, to give his humor a savvy edge. And language plays a big part in the wit and humor equation; El Paso's spicy salsas and marinades include Smoke Signals, Snake Bite, and Hell Fire Damnation. Playful drawings wittily represent the ingredients, while signaling a sophisticated aura for each product. These are not merely

crass commercial trade characters, like the Frito Bandito, but a lexicon of graphic symbols with unique semiotic powers. Animated pasta, pretzels and cashews join hands and joyfully dance around blazing logs in the Pow Wow Trail Mix while a personable Prince of Orange stares with charm from the top of his cake box.

El Paso Chili Company has a concerted humor strategy, while other products are imbued with wit when the designer views it as appropriate to the nature of the product. Haley Johnson's Palais d'Amour honey jars are both charming and fun, and position this honey apart from other, serious approaches. At first glance, the head of a woman with an eighteenth-century coif complements the product's nod to royalty in its name, yet on closer perusal bees fly in and out of her hair underscoring the term beehive hairdo. This delightful mascot in a surreal context is at once an appealing messenger of the product's virtues and a subtle jab at the conventions of food design.

Humorous packages for common products are effective when they happily surprise the consumer, which is the essence of impulse or gift consumption. But to have sustained success the combination of package and product must maintain a level of interest that is impossible to accomplish with humor alone. While many of the products in this section have brand loyal customers, the hook may be wit but the ultimate test is quality.

Appetizing Companions

Client
Ham I Am

Firm
SullivanPerkins

Art Director
Ron Sullivan

Designers/Illustrators
Clark Richardson
Art Garcia

Photographer
Gerry Kano

Palais Dàmour
Honeymoon Sweet

Client
Palais D'Amour Honey

Firm
Haley Johnson Design Co.

Designer/Illustrator
Haley Johnson

Photographer
Paul Irmiter,
Irmiter Photography

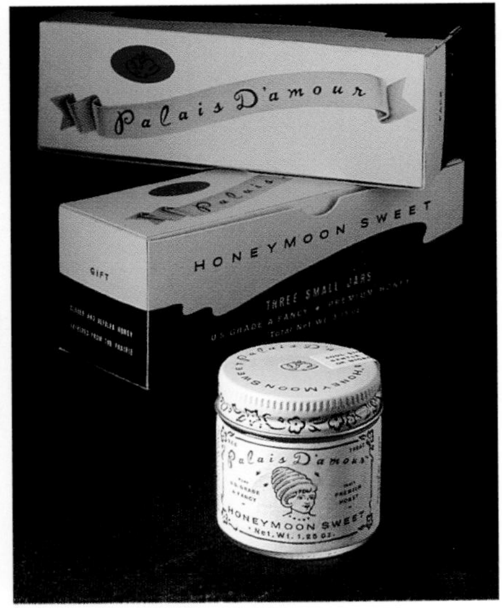

In Cahoots

Client
In Cahoots

Firm
Werner Design
Werks, Inc.

Art Director/
Designer/Illustrator
Sharon Werner

Photographer
Dave Bausman

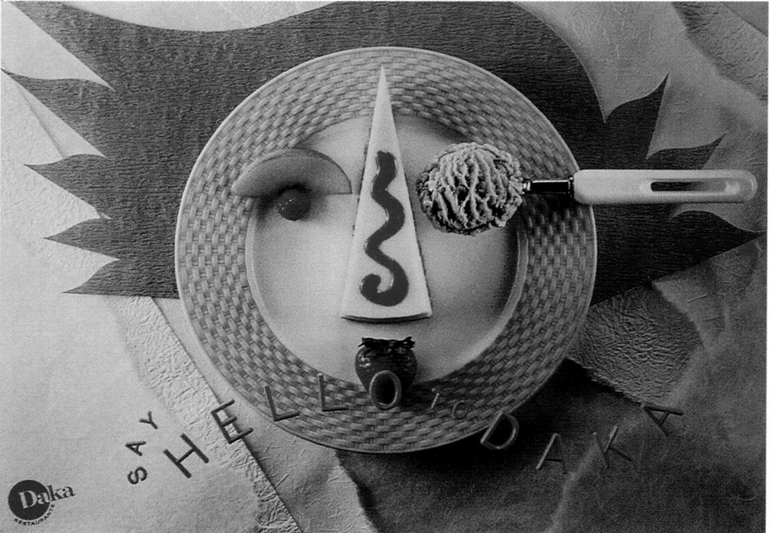

Let's Do Lunch
Posters and Postcards

Client
Daka International

Firm
SullivanPerkins

Art Director/Designer
Art Garcia

Photographer
Robb Debenport

Christmas Cookie Card

Client/Firm
SullivanPerkins

Art Director
Ron Sullivan

Designer
Jon Flaming

Photographer
Tim Boole

Mac Andrew's Scotch Ale

Client
Merchant du Vin
Corporation

Art Director/Designer
Charles Finkel

Numero Uno
Margarita Mix

Client
El Paso Chile Company

Firm
RBMM

Art Director/Designer
Luis Acevedo

Illustrator
Wayne Jonson

It's Hard to
Have a Bad Day

Client/Firm
Shoebox Greetings

Art Directors
Karen Brunke
Julie McFarland

Designer/Illustrator
Meg Cundiff

Copywriter
Deanne Stewart

Elephants Leap

Client
Caprock Winery

Firm
SullivanPerkins

Art Director/Designer
Kelly Allen

Photographer
Gerry Kano

OK Cola & Collateral

Client
The Coca-Cola Company

Firm
Wieden & Kennedy

Creative Director/
Art Director/Designer
Todd Waterbury

Writer
Peter Wegner

Illustrators
Calef Brown
Charles Burns
David Cowles
Daniel Clowes

Photographer
Mark Ebsen

OK BRAND SODA
OK.
A CARBONATED "BEVERAGE"
"OK" GLASSES

THE "OK" MANIFESTO
1. WHAT'S THE POINT OF "OK"? WELL, WHAT'S THE POINT OF ANYTHING?
2. "OK" SODA EMPHATICALLY REJECTS ANYTHING THAT IS NOT OK, AND FULLY SUPPORTS ANYTHING THAT IS.
3. THE BETTER YOU UNDERSTAND SOMETHING, THE MORE OK IT TURNS OUT TO BE.
4. "OK" SODA SAYS, "DON'T BE FOOLED INTO THINKING THERE HAS TO BE A REASON FOR EVERYTHING."
5. "OK" SODA REVEALS THE SURPRISING TRUTH ABOUT PEOPLE AND SITUATIONS.
(OVER)
OK
A CARBONATED "BEVERAGE"

OK®
"OK-NESS"
ORDINARY/SPECIAL
OPTIMISTIC/IRONIC
"BEVERAGE"
"BEVERAGE"
"BEVERAGE"
"BEVERAGE"
OK.
OK.
OK BRAND SODA

Tropical Source Organic Bars

Client
Cloud Nine, Inc.

Firm
Haley Johnson Design Co.

Designer/Illustrator
Haley Johnson

Photographer
Paul Irmiter,
Irmiter Photography

Dry Ice Seltzer

Client
Lombardi Dry Ice Seltzer

Firm
Madame Sophie's

Art Director/Designer
Karen E. Burgess

Mangia Pizza

Client
Mangia—Chicago Stuffed Pizza

Firm
Hixo

Art Directors
Mike Hicks
Duana Gill

Designer
Duana Gill

Yahoo Bar-B-Q

Client
Ham I Am

Firm
SullivanPerkins

Art Director/ Designer/Illustrator
Kelly Allen

Photographer
Gerry Kano

Boston Chocolate

Client
Boston Chocolate Co.

Firm
Werner Design Werks, Inc.

Art Director/
Designer/Illustrator
Sharon Werner

Photographer
Paul Irmiter

Bone Dry Beer

Client
Bone Dry Beer
Microbrewery

Firm
Margo Chase Design

Art Director
Margo Chase

Designers
Wendy Ferris
Anne Burdick

Photographer
Sidney Cooper

Bubba Brand Coffee
and Heatin' Hot Sauce

Client
Atlantis Coastal Foods

Firm
Gil Shuler Graphic
Design

Art Director
Gil Shuler

Designer
Jay Parker
Gil Shuler

Salsa Divino

Client
El Paso Chile Company

Firm
Anderson Design

Art Director/
Designer/Illustrator
Charles Anderson

Pow Wow Trail Mix

Client
El Paso Chile Company

Firm
Pirtle Design

Art Director/Designer
Leslie Pirtle

Illustrator
Lisa Haney

Buffalo Bill's Ales

Client/Firm
Buffalo Bill's Brewery

Art Director
Bill Owens

Designer
Karen Barry

Carolina Swamp Stuff

Client/Firm
Carolina Swamp Stuff

Designer/Photographer
Mary Ellen Box

Nantucket Seasonings

Client
Nantucket Offshore
Seasoning

Coyote Nuts

Client
El Paso Chile Company

Firms
Borland Design
El Paso Chile Company

Art Director
Park Kerr

Designer/Illustrator
Chaz Borland

Salsa Primera
Chile Con Queso
Mango Tango

Client
El Paso Chile Company

Firm
Pirtle Design

Art Director/Designer
Leslie Pirtle

Illustrator
Lisa Haney

Buffalo Bill's Ales

Client/Firm
Buffalo Bill's Brewery

Art Director
Bill Owens

Designer
Karen Barry

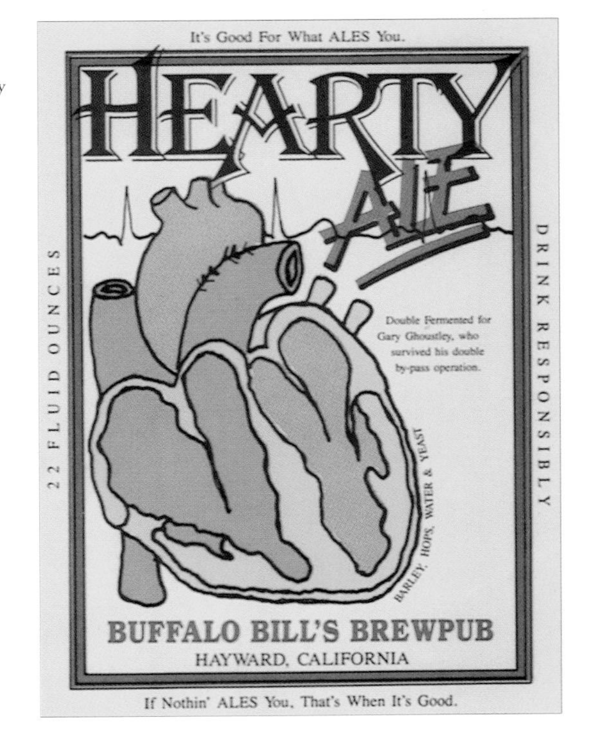

Chipotle Cha Cha Cha

Client
El Paso Chile Company

Firms
Pirtle Design
El Paso Chile Company

Art Director
Park Kerr

Designer
Leslie Pirtle

Illustrator
Lisa Haney

Hellfire & Damnation

Client
El Paso Chile Company

Firm
Pirtle Design

Art Director
Leslie Pirtle

Designers
Leslie Pirtle
Woody Pirtle

Illustrator
Woody Pirtle

Cinco de Mayo

Client
El Paso Chile Company

Firms
Pirtle Design
El Paso Chile Company

Designer
Leslie Pirtle

Illustrators
Leslie Pirtle
Woody Pirtle

Clyde's Chili

Client
Britches of Georgetown

Firm
Grafik
Communications Ltd.

Creative Director
Judy Kirpich

Designers
Melanie Bass
Judy Kirpich

Photographer
David Sharpe

Desert Pepper Bean Dip

Client
Desert Pepper Trading Co.

Firm
Michael Mabry Design Inc.

Designer/Illustrators
Michael Mabry

Photographer
Michael Lamotte

My Pig-Out Food

Client/Firm
Shoebox Greetings

Art Directors
John Wagner
Julie McFarland

Designer/Illustrator
Meg Cundiff

Cafe Marimba Salsa

Client
Cafe Marimba

Firm
Michael Mabry
Design Inc.

Photographer
Michael Lamotte

Zélé Panaché

Client
Seh Importers

Firm
Michael Mabry
Design Inc.

Designer/Illustrator
Michael Mabry

Photographer
Michael Lamotte

Prince of Orange
Bundt Cake

Client
El Paso Chile
Company.

Firm
Michael Mabry
Design Inc.

Designer/Illustrator
Michael Mabry

Photographer
Michael Lamotte

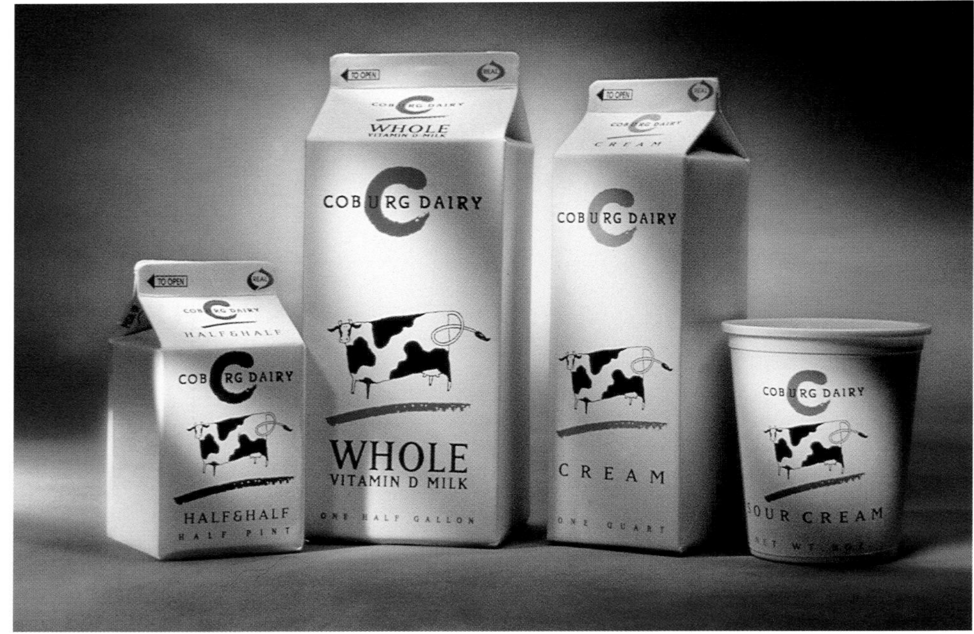

Coburg Dairy

Client
Coburg Dairy, Inc.

Firm
Pedersen Gesk

Art Director
Sig Gesk

Designer
Tjody Overson-de-Vaal

Photographer
Courtesy of Pedersen Gesk

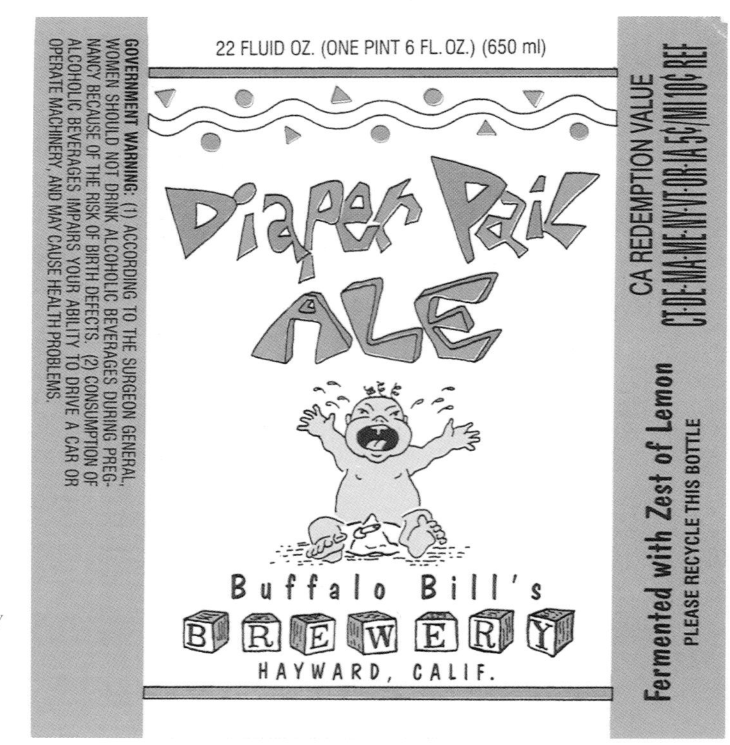

Buffalo Bill's Ales

Client/Firm
Buffalo Bill's Brewery

Art Director
Bill Owens

Designer
Karen Barry

Firehog Chili

Client/Firm
Hixo

Art Directors
Tom Poth
Mike Hicks

Designer
Mike Hicks

Club Iowa Lemon Bite

Client
Grand Palace Foods
International

Firm
Supon Design Group

Creative/Art Director
Supon Phornirunlit

Designer/Illustrator
David Carroll

Photographer
Oi Jakrarat Veerasarn

Inner Beauty Sauce

Client/Firm
Inner Beauty/
Last Resort Inc.

Art Director
Chris Schlesinger

Designer/Illustrator
Michael W. Buckley

Oh Natural

Don't go [food] shopping when you're hungry, goes the old bromide. But this good advice is rarely heeded, and food manufacturers know it. They already use every conceivable tactic to lure consumers into buying their products, so being hungry only makes it harder to resist the strong allure. This is key to the concept of appetite appeal. If a customer is in a suggestive frame of mind to begin with, the hypnotic lure of a tasty photograph or drawing of the product itself—not some abstract depiction or typographic design—will connect with the appetite center of the brain, which signals the motor function to grab the box, bottle, or tin off the shelf. Tapping into this power of suggestion is why many packages show the typical bite taken out of a piece cake, spoon strategically positioned in a bowl of steamy soup, and fork poised to attack a plate of spaghetti. As further mouthwatering bait, ever more frequently today, the product is shown au natural *behind cellophane wrappers or in clear bottles and jars.*

The supermarket is purposefully designed to project a variety of design enhancements, and hook the shopper. Bright fluorescent lights illuminate the stark color schemes on many packages, but the typical primaries —reds, greens, yellows and blues—which jump off the shelves are such a common conceit that they tend to blend into a grand chromatic mishmash. So as designers are confronted with increasingly greater difficulties in setting products apart, they are turning to the products themselves for inspiration. In a sea of pictorial and typographic packaging, the clear container has a greater chance of gaining attention.

Au natural is not, however, appropriate for every product. Those with a certain eye appeal, like herbal vinegars with the herbs and vegetables swimming in liquid, gain the most benefit from this method. Yet the concept began not as a marketing tactic but developed over a century ago as a functional way to package homemade preserved and pickled products. Eventually it became a means to convey and sell merchandise produced by cottage industries. Take an old mason jar, cover it with a swatch of gingham, and affix a handwritten mailing label, and presto—a package. With the increase of small homemade food businesses this method of packaging signaled a variety of virtues, not the least of which was the fact that it is an alternative to the mass-produced, assembly line food package.

In recent years the look-and-see container has been adopted by larger food manufacturers for those products claimed to be pure and natural. In addition to pragmatics, this kind of package affirms certain personal, social, and environmental concerns. Since glass bottles and jars are recyclable, consumers believe that choosing this type over a more wasteful package has its own inherent virtue. Moreover, the au natural container implies that the contents are truly natural ingredients. In an age when "natural" has become a vague, if banal, marketing term, seeing is believing carries a lot more weight than a typographic assertion on the label.

Au natural packages are not all made of see-through glass. This genre also includes packages made from earth-friendly materials. Companies with environmental sensitivity have found a variety of additional recycled containers that both telegraph their concerns and present the consumer with effective eye-catching packages. The Seattle firm, Hornall Anderson claims to have created the identity for the Starbucks line with such heightened world issues in mind. Their goal was to design each store from signage to containers with a style that was at once consumer friendly and environmentally thrifty. Starbuck's coffee bags are made from kraft paper which has been clay coated, both ensuring freshness and limiting waste.

In the honest and fashionable push towards natural ingredients and natural packaging, illustrative graphics have also markedly changed from previous hardsell conventions to emphasize the softer sell of natural environments. Pastel colors, naturalistic settings, and rain forest references have become the new clichés of the Oh Natural generation.

Sparkling Ice

Client
Talking Rain

Firm
Hornall Anderson
Design Works

Art Director
Jack Anderson

Designers
Jack Anderson
Julia LaPine
Jill Bustamante
Heidi Favour
Jana Nishi
Leo Raymundo

Illustrator
Julia LaPine

Photographers
Darrell Peterson (*bottles*)
Tom McMackin (*cans*)

Contains 5% Fruit Juice
berry
SPARKLING
ICE
12 FL OZ (355 ml)
apple
SPARKLING
ICE
APPLE FLAVORED
cherry
SPARKLING
ICE
12 FL OZ (355 ml)
peach
SPARKLING
ICE
PEACH FLAVORED
SPARKLING WATER WITH
12 FL OZ (355 ml)

Empress International Seafood Packaging

Client
Empress International Ltd.

Firm
Pentagram Design

Art Director
Paula Scher

Designers
Paula Scher
Ron Louie

Illustrator
Douglas Smith

Photographer
Barry Robinson

Veg Out Packaging

Client
Silverado Foods

Firm
Pentagram Design

Art Director
Paula Scher

Designers
Paula Scher
Ron Louie

Photographer
John Paul Endress

Biale Wine

Client
Robert Biale Vineyards

Firm
Colonna Farrell Design

Art Director
Cynthia Maguire

Designer
Chris Mathes Baldwin

Illustrator
Mike Gray

Photographer
David Bishop/
San Francisco

Coffee

Client
Coffee Plantation

Firm
Estudio Ray

Art Directors
Christine Ray
Joe Ray

Designers
Christine Ray
Joe Ray
Leslie Link

Illustrator
Joe Ray
Frank Ybarra

Photographer
Michael Fioritto

Spa Mineral Water
Client
Spa Monopole Corp.

Spa Mineral Water
Client
Spa Monopole Corp.

Messinia
Olive Oil

Client
Classical Foods

Firm
Louise Fili Ltd.

Art Director/ Designer
Louise Fili

Photographer
Ed Spiro

Canyon Ranch
Spa Cuisine

Client
Food Development & Marketing, Inc.

Firm
Tana & Co.

Art Director/Designer
Tana Kamine

Illustrator
Anthony Russo

Photographer
Thom DeSanto

Peaberry Coffee

Client
Peaberry Coffee L

Firm
VSA Partners, Inc

Art Director/Design
James Koval

Illustrator
Mary Frock Lemp

Photographer
Glen Gyssler

PEABERRY
COFFEE
TM
FRENCH ROAST
LIGHT
PEABERRY
COFFEE
TM

THE ESSENTIAL MASALA
THE MINI INDIAN SPICE KITCHEN
In Hindi, "Masala" means mixture. "The Essential Masala" is a mixture of ingredients often difficult to find outside of India. Each ingredient adds an authentic note to Indian dishes and together they bring magic to everyday cooking
Culinary Alchemy
COOKING WITH THE SPICES OF
India
Easy to Follow Recipes to Transform Ordinary Foods into Extraordinary Meals
THE DRY TRINITY
THE INDIAN SPICE KITCHEN
CORIANDER
TURMERIC
CUMIN
For Sharabi Kababis Everywhere
THE ESSENTIAL MASALA
FENUGREEK
AMCHOOR
ASAFETIDA
AJWAIN
PANCH PHORAN
CLOVES
BLACK CUMIN
URAD DAL
KALONJI
Mustard
GARAM MASALA
NET WT. 5oz
NET WT. 3.95oz
TAMARIND
MASOOR & MOONG DAL
NET WT. 7.25oz 203g
THE EXOTICS
Includes 6 Recipes
THE INDIAN SPICE KITCHEN
ASAFETIDA
BLACK CUMIN
FENUGREEK
CLOVES
AJWAIN
URAD DAL
AMCHOOR

M.A. O'Halloran
Shortbread

Client
M.A. O'Halloran
A Family Baking
Company

Photographer
Holly Stewart

he Essential Masala

lient
ulinary Alchemy

irm
he Office of
ichael Manwaring

rt Director
athleen O'Rourke

esigners
lizabeth Manwaring

hotographer
ohn Clayton

Capons Rotisserie Chicken
Identity Program

Client
Capons Rotisserie Chicken

Firm
Hornall Anderson
Design Works

Art Director
Jack Anderson

Designers
Jack Anderson
David Bates
Cliff Chung

Illustrators
David Bates
George Tanagi

Photographer
Tom McMackin

CAPONS
ROTISSERIE CHICKEN
CAPONS
ROTISSERIE CHICKEN
CAPONS
ROTISSERIE CHICKEN

FROM OUR ROTISSERIE
SIGNATURE SAUCES
FROM OUR OVENS
SALADS
SANDWICHES
SPECIAL COMBINATIONS
QUARTER CHICKEN & SALAD 5.00
HALF SANDWICH & SALAD 5.25
HALF SANDWICH, SOUP & SALAD 5.95
COMPLETE FAMILY DINNERS 19.95
FAMILY ROTISSERIE CHICKEN DINNER
FAMILY CHICKEN LASAGNA DINNER
SOUPS
CAPONS CHICKEN NOODLE 1.25 2.35
VEGETABLE MINESTRONE 1.25 2.35
FOR KIDS 12 & UNDER
ROTISSERIE CHICKEN DINNER 2.95
GRILLED CHEESE & FRUIT 2.75
ALPHABET SOUP 1.35
SIDES & EXTRAS
Signature Sauce (4 oz.) .50
Cornbread Muffin .40
Dinner Sides (6 oz.) 1.25
DESSERTS
CARAMEL APPLE CRUMBLE 1.95
OATMEAL FRUIT BAR 1.00
CREAM CHEESE BROWNIE 1.00
COLD BEVERAGES
Soft Drinks .85
Bottled Juices .85
Iced Tea
Fresh Squeezed Lemonade
BEER & WINE
Assorted Micro Brews
Cabernet Sauvignon or Chardonnay
HOT BEVERAGES
Coffee
Assorted Teas
Hot Chocolate
ESPRESSO
Latte
Mocha
Extra Shot .40
Flavors

Club Iowa
Sassy Strawberry

Client
Grand Palace Foods International

Firm
Supon Design Group

Creative/Art Director
Supon Phornirunlit

Designer/Illustrator
David Carroll

Photographer
Oi Jakrarat Veerasarn

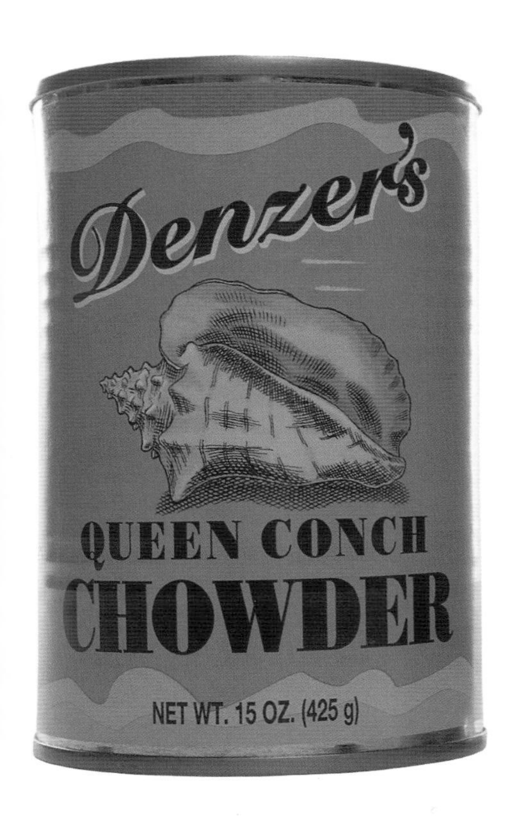

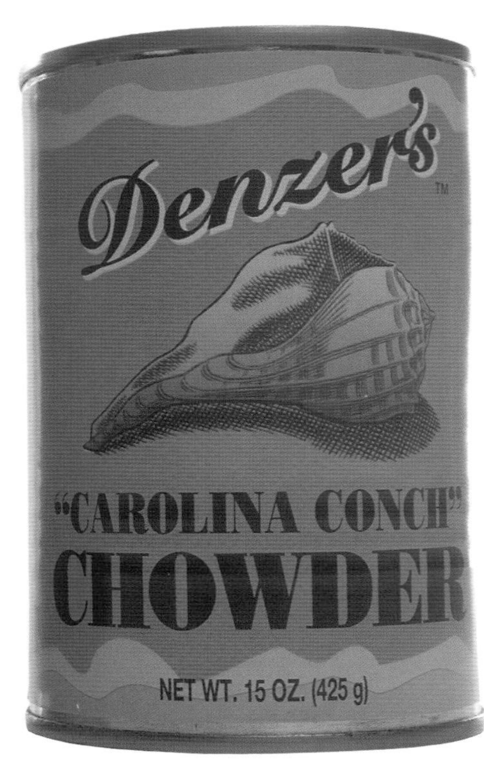

Denzer's Chowders

Client
Denzer's Food Products

Firms
Found Graphic Design

Piedmont Label Co.

Art Director
Nina Tou

Illustrator
David Nelson

Stretch Island
Fruit Leather

Client
Stretch Island

Firm
Hornall Anderson
Design Works

Art Director
Jack Anderson

Designers
Jack Anderson
Mary Hermes
Heidi Favour

Illustrator
Fran O'Neill

Photographer
Tom McMackin

MALTED WHEAT
BROADMOOR
GRAIN & SEED
GOLDEN WHEAT
OAT BRAN

BROADMOOR BAKER
BROADMOOR
BAKER
BROADMOOR
BAKER

BROADMOOR
BAKER
OAT BRAN
GRAIN & SEED
GOLDEN WHEAT

Broadmoor Baker
Packaging & Collateral

Client
Broadmoor Baker

Firm
Hornall Anderson
Design Works

Art Director
Jack Anderson

Designers
Jack Anderson
Mary Hermes
Jani Drewfs

Illustrator
Scott McDougall

Photographer
Tom McMackin

Crystal Geyser Sparkling Mineral Water

Client
Crystal Geyser

Firm
Glenn Martinez and Associates

Art Director/Designer
Glenn Martinez

Illustrator
Steve Doty

Photographer
Don Silverek

San Anselmo's Biscotti Super Naturals

Client
San Anselmo's Biscotti

Firm
Sharon Till Associates

Art Director/Designer
Sharon Till

Illustrator
Rik Olson

Photographer
Matt Farruggio

The Ojai Cook
Taste of California

Client
Sweet Adelaide
Enterprises, Inc.

Firm
Chris May Design

Art Directors
Chris May
Joan Vogel

Designer
Chris May

Illustrator
Jennifer Kirk

Photographer
Chuck Gebhart

ADONIS
SPRING WATER
12 FL. OZ. Non-carbonated Sodium Free

Fresh Market

Client
Grand Palace Foods International

Firm
Supon Design Group

Creative/Art Director
Supon Phornirunlit

Designer
Apisak "Eddie" Saibua

Photographer
Oi Jakrarat Veerasarn

nis Spring Water

nt
nd Palace
ds International

n
on Design Group

tive Director
on Phornirunlit

Directors
on Phornirunlit
rew Dolan

gner/Illustrator
rew Dolan

tographer
akrarat Veerasarn

MRS. POTTS' TEA

A CUP OF COMFORT

Fine Black Tea Blended with Berries

CAUGHT WITHOUT YOUR POT? USE THE REPUBLIC'S ROUND UNBLEACHED TEA BAG.

1. Fill the kettle with fresh, cold water and then heat to boiling.

2. Place one tea bag at the bottom of your cup.

3. When the water reaches a rolling boil, pour the water over the tea bag and infuse. Taste the tea after one minute. When the tea tastes right to you, remove the tea bag. Enjoy sip by sip.

CAFFEINE CONTENT
milligrams per 5 ounce cup

Coffee
Black Tea
Oolong Tea
Green Tea

Mrs. Potts' artwork from an Ann Hould-Ward costume design.

The REPUBLIC of TEA

brews 50 cups

Natural Unbleached Tea Bags

NET WT 3oz (85g)

The Beast's well-meaning servants are all under the same curse as their master, through no fault of their own. Among them, the motherly and endearing cook, Mrs. Potts, has been transformed into a teapot, and her son Chip is a lively teacup. With every passing day, the Enchanted Objects are changing a bit more. Unless Belle helps break the spell, they must give up hope of becoming human again.

A Sweet, Enchanting Brew

FINE ORGANIC ASSAM TEA IS BLENDED WITH NATURALLY SWEET BLACKBERRY LEAVES AND JUICY BLUEBERRIES TO CREATE MRS. POTTS' FAVORITE TEA. SERVE IT HOT FOR A SOOTHING, TRADITIONAL CUP OR OVER ICE FOR A FRESH ALTERNATIVE. THIS TEA ALWAYS TASTES BEST SERVED IN THE COMPANY OF FRIENDS.

BEAUTY AND THE BEAST
THE BROADWAY MUSICAL

INGREDIENTS *Organic Assam black tea certified by The Soil Association of the U.K., blackberry leaves, blueberries, and natural flavors*

This container is recyclable and reusable in the home

7 42676 10085 0

TEAS FOR ALL SEASONS

A Quartet of Full-Leaf Teas and Herbs

AUTUMN

CINNAMON HERB TEA

DRINKING TEA BEGINS WITH BREWING TEA

1. Fill the kettle with fresh, cold water and then heat.

2. When the water reaches a rolling boil, warm the teapot by swirling steamy water in it and then pour it out.

3. Place one level teaspoon of leaves per cup into your pot or infusing basket.

4. Pour the water over the tea leaves and infuse for two to five minutes. Experiment to find your favored brewing time.

5. Pour the tea. Avoid letting the leaves stew. Remove the infusing basket or pour the remainder of the tea into a thermos to enjoy later.

CAFFEINE CONTENT
milligrams per 5 ounce cup

Herbal Tea
Green Tea
Oolong Tea
Black Tea
Coffee

The REPUBLIC of TEA

NET WT 3.5oz (100g)

Pulling up a favorite chair to the hearth-side, cup of steaming hot tea in hand, is one reward enjoyed during the harvest season. After days in the orchard gathering apples or picking grapes in the vineyard, the soothing aromatic brew relaxes tired muscles. Fields of ripe wheat, bundles of drying corn, cattails in the marsh, cellars filling with the bounty of harvest, all reflect preparation for the approaching season. Geese flying in formation across the full harvest moon foreshadow this change.

Autumn Herbs and Spices

AN EXOTIC HERBAL BLEND MADE FROM A RARE, GRASSY HERB OF SOUTH AFRICA CALLED ROOIBOSCH (PRONOUNCED "ROYBOSH".) "ROI" TEA OFFERS A BEAUTIFUL CARAMEL-COLORED LIQUOR, A DELICATE AND SMOOTH TASTE WHICH WE ENLIVEN WITH A DASH OF CINNAMON AND OTHER WARMING SPICES THAT WILL TAKE THE EDGE OFF A COOL FALL DAY. NATURALLY CAFFEINE FREE.

Our "seasonal" teapot was painted by California artist Maryjo Koch. This watercolor and gouache illustration comes from her book **Delectables for All Seasons: Tea**, produced for Swans Island by Collins Publishers/ San Francisco.

INGREDIENTS *Rooibosch, cinnamon, natural flavors*

This container is recyclable and reusable in the home

For more information on our tea and teaware write to The Minister of Supply: P.O. Box 1175 Mill Valley, CA 94942

7 42676 10073 7

The Republic of Tea Labels

Client/Firm
The Republic of Tea

Art Director
Nancy Bauch

Designers
Nancy Bauch
Vic Zauderer

Illustrators
Patricia Ziegler
Faye Rosenzweig
Gina Amador

TEAS FOR ALL SEASONS

A Quartet of Full-Leaf Teas and Herbs

WINTER

BLACK TEA WITH IMPERIAL SPICES

DRINKING TEA BEGINS WITH BREWING TEA

1. Fill the kettle with fresh, cold water and then heat.

2. When the water reaches a rolling boil, warm the teapot by swirling steamy water in it and then pour it out.

3. Place one level teaspoon of leaves per cup into your pot or infusing basket.

4. Pour the water over the tea leaves and infuse for two to five minutes. Experiment to find your favored brewing time.

5. Pour the tea. Avoid letting the leaves stew. Remove the infusing basket or pour the remainder of the tea into a thermos to enjoy later.

CAFFEINE CONTENT
milligrams per 5 ounce cup

Green Tea
Oolong Tea
Black Tea
Coffee

The REPUBLIC of TEA

NET WT 3.5oz (100g)

Winter allows for both solitude and for celebration in the company of friends. It is a time to reflect upon the seasons past and to rekindle our hopes, dreams, and faith in the season to come. Boughs of pine and fir adorn both the mantle and holiday table. Festive candles illuminate the feast prepared from favorite traditional recipes awaiting invited guests. The spicy fragrance pouring forth from under the tea cozy signals of the pleasures to come. In the quiet moon-lit stillness at the forest's edge, a solitary deer steps cautiously onto the fragile crust of the sparkling snow.

Imperial Spices

CELEBRATE THE HOLIDAYS — OR THE WINTER SOLSTICE — WITH THIS FINE BLACK TEA, BLENDED WITH TRADITIONAL IMPERIAL SPICES INCLUDING CLOVES, NUTMEG, AND CINNAMON. FINE TEA LEAVES GROWN IN SRI LANKA, CHERISHED FOR THEIR LIGHT CUP, ARE ACCENTUATED WITH FRESH GROUND SPICES. LOVELY TO SIP, LOVELY TO SMELL — THE SWEET AND SPICY AROMA WILL FILL YOUR HOME WITH A FESTIVE SPIRIT.

Our "seasonal" teapot was painted by California artist Maryjo Koch. This watercolor and gouache illustration comes from her book **Delectables for All Seasons: Tea**, produced for Swans Island by Collins Publishers/ San Francisco.

INGREDIENTS *Fine black tea, cloves, cinnamon, orange peel*

This container is recyclable and reusable in the home

For more information on our tea and teaware write to The Minister of Supply: P.O. Box 1175 Mill Valley, CA 94942

7 42676 10070 6

SIP BY SIP
The REPUBLIC of TEA™
NOT GULP BY GULP

Oh Natural

The Republic of Tea
Packaging & Labels

Client/Firm
The Republic of Tea

Art Director
Nancy Bauch

Designers
Nancy Bauch
Vic Zauderer

Illustrators
Patricia Ziegler
Faye Rosenzweig
Gina Amador

Photographer
Mario Parnell

Tea of Good Tidings

Winter Fruit Blend

Finest Quality Black Tea Blended with Fruit and Spices

brews 50-60 cups
Full-Leaf Loose Tea
NetWt 3.5oz (100g)

Drinking tea begins with Brewing Tea

1. Fill the kettle with fresh, cold water and then heat.
2. When the water reaches a rolling boil, warm the teapot by swirling steamy water in it and then pour it out.
3. Place one level teaspoon of leaves per cup into your pot or infusing basket.
4. Pour the water over the tea leaves and infuse for two to five minutes. Experiment to find your favored brewing time.
5. Pour the tea. Avoid letting the leaves stew. Remove the infusing basket or pour the remainder of the tea into a thermos to enjoy later.
6. Add the used tea leaves to your garden. They make a healthy fertilizer for flowering plants.

Caffeine Content
milligrams per 5 ounce cup
Green Tea
Oolong Tea
• Black Tea
Coffee

The Gift of the Leaf represents many virtues; in Japan, it is an offering of peace. In China, it is a wish for well-being. In Europe, it is a time for civility. In The Republic, it is cause to appreciate the moment and live life sip by sip, rather than gulp by gulp.

'Tis the Season

A cornucopia of winter harvested fruit and spices is blended with fine black tea leaves. Whole cranberries and red orange peel offer a sweet fruitiness. Cinnamon and juniper berries lend a spicy finish. Sip a cup and capture the holiday spirit.

Our nature-inspired teaware enhances the preparation and enjoyment of full-leaf teas.
For a catalogue, write:
The Minister of Supply (020)
Post Office Box 1589
Novato, California 94948-1589
Comments? Call: 1-800-298-4TEA

Ingredients *Fine black tea, juniper berries, natural flavors, cranberries, orange peels, cloves, vanilla, rose petals, almond pieces, black currants, blackberry leaves, cardamon*

This container is recyclable and reusable.

Sip by Sip rather than gulp by gulp

© 1995 The Republic of Tea
Novato, California 94949

7 42676 10100 0

Desert Sage Tea

Clear the Mind Herb Tea

Naturally Caffeine Free

The Republic of Tea

brews 60 cups
Fresh Full-Leaf Herbs
NetWt 3oz (85g)

Drinking tea begins with Brewing Tea

1. Fill the kettle with fresh, cold water and then heat.
2. When the water reaches a rolling boil, warm the teapot by swirling steamy water in it and then pour it out. Herbs infuse best in a glass or ceramic pot.
3. Place one level teaspoon of herbs per cup into your pot or infusing basket.
4. Pour the water over the herbs, cover, and infuse for five or more minutes.
5. Pour the tea. Avoid letting the herbs stew. Remove the infusing basket or pour the remainder of the tea into a thermos to enjoy later.

Caffeine Content
milligrams per 5 ounce cup
• Herbal Tea none
Green Tea
Oolong Tea
Black Tea
Coffee

Where there is little rain, plants do not develop the extravagant flowers and foliage of their cousins in the wetter bioregions. Instead, their energy goes into aromatic oils, which help defend the sparse-leafed plants against predators. Distinct-smelling, strong-tasting herbs such as white sage and desert tea often look like "weeds," but reveal their beneficial essence when crushed, burned or steeped in boiling water.

The Minister of Herbs

Breathe Deeply

The heady aroma of white sage evokes images of the high southwest. To this herb we've added full-bodied rooibosch and cool mint notes for a satisfying and complex brew. Drink it with a little honey on a cold, drizzly day and feel the dry desert heat radiate to your core.

Our nature-inspired teawares enhance the preparation and enjoyment of full-leaf teas.
Write for a Catalog
to The Minister of Supply
Post Office Box 1175
Mill Valley, California 94942 USA

Ingredients *Rooibosch, orange bergamot mint, desert tea, blackberry, white sage*

This container is recyclable and reusable in the home

© 1994 The Republic of Tea, Inc.
Novato, California 94949

Chamomile Lemon

Surrender to Sleep Herb Tea

Naturally Caffeine Free

brews 60 cups
Fresh Full-Leaf Herbs and Flowers
NetWt 1.75oz (50g)

Drinking tea begins with Brewing Tea

1. Fill the kettle with fresh, cold water and then heat.
2. When the water reaches a rolling boil, warm the teapot by swirling steamy water in it and then pour it out. Herbs infuse best in a glass or ceramic pot.
3. Place one level teaspoon of herbs per cup into your pot or infusing basket.
4. Pour the water over the herbs, cover, and infuse for five to ten minutes. Experiment to find your favored brewing time.
5. Pour the tea. Remove the infusing basket or pour the remainder of the tea into a thermos to enjoy later. This tea is also makes a relaxing cooler over ice.

Caffeine Content
milligrams per 5 ounce cup
Coffee
Black Tea
Oolong Tea
Green Tea
• Herbal Tea none

The leaves take the water, and the tea takes me. I put up no resistance, surrendering myself to what has entered. Soon I am the tea, and the tea is me.

The Minister of Leaves

Surrender to Sleep

Egyptian chamomile is blended with organic lemon balm, (grown for us in the Pacific Northwest) to produce a fragrant, soothing cup. These and other potent botanicals surrender a most sweet and tranquil tea, delighting the palette, body, and mind. Nice before bed.

Our nature-inspired teawares enhance the preparation and enjoyment of full-leaf teas.
Free Catalog
Write: The Minister of Supply
Post Office Box 1175
Mill Valley, California 94942 USA
415 721 2177

Ingredients
Lemon balm, linden flower, orange blossoms, skullcap, chamomile, lavender flower, passion flower, valerian root

This container is recyclable and reusable in the home

© 1994 The Republic of Tea Inc.
Novato, CA 94949

Aleatico

Client
Sutter Home

Firm
Michael Osborne
Design

Art Director
Michael Osborne

Designer/Illustrator/
Typographer
Tom Kamegai

Hogan's Market
Grocery Bags

Client
Hogan's Market/
Puget Sound Marketing
Corporation

Firm
Hornall Anderson
Design Works

Art Director
Jack Anderson

Designers
Jack Anderson
Julia LaPine
Denise Weir
Lian Ng

Illustrator
Larry Jost

Typographer
Nancy Stenz

Photographer
Tom McMackin

African Bread Mix

Client
McCleary & Co. Inc.

Firms
Pac National Inc.
McCleary & Co. Inc.

Designers
Glenda McCleary
James Dorn

Pororoca

Client
Plaza Design

Firm
Sussman/Prejza
& Co., Inc.

Architect
Fernando Vazquez

Designer
Deborah Sussman

Photographer
Annette del Zoppo
Productions

Toucan Chocolates

Client
Toucan Chocolates

Firms
Egg Design Partners
Imprimatur

Art Director
Michael Sullivan

Designers
Cynthia Delfino
Jonathan Jackson

Illustrator
Nancy Nimoy
Polly Becker

TOUCAN CHOCOLATES
TOPAZ TOFFEE
Crunchy Brazil nut toffee coated with chocolate and cashews
Net Wt. 6 oz. (170g)
TOUCAN CHOCOLATES
Net Wt. 2¼ oz. (63g)

Georgio Mundelli's
Premiere Catsup

Client
George Mundell, III

Firm
Fortune Design Studio

Art Director
Lani Fortune

Designers
John Fortune
Lani Fortune

Illustrator
John Fortune

Photographer
Joel Levin Photography

Twin Valley Popcorn & Collateral

Client
Twin Valley Popcorn

Firm
Love Packaging Group

Art Director/ Designer/Illustrator
Tracy Holdeman

Photographer
Don Siedhoff

Paula's Salsa Novas

Client
Sweet Adelaide

Firm
Kimberly Baer
Design Associates

Art Director
Kimberly Baer

Designer
Barbara Cooper

Illustrator
Nancy Nimoy

Photographer
Chuck Gebhart

Arizona Chocolate
Flavored Drinks

Client
Arizona Beverages

Photographer
Bill Streicher

Equatorial Provisions

Client
Chocolate Moon

Firm
Creative Forces

Designer
Clare Ahern

Fajita & Cajun Dust

Client
Allen & Cowley Urban
Trading Co.

Nantucket Nectars

Client
Nantucket Nectars

Firm
Brand Equity
International

Photographer
Jefferson White

2002 Carbonated
Fruit Brew

Client
West End Products

Firm
Zipatoni

Art Director/Designer
Tom Corcoran

Illustrator
Lingta Kung

Photographer
Marty Keeven

Napa Ridge
Case Cards

Client
Wine World
Estates

Firm
Halleck Design
Group

Designer
Bob Dinetz

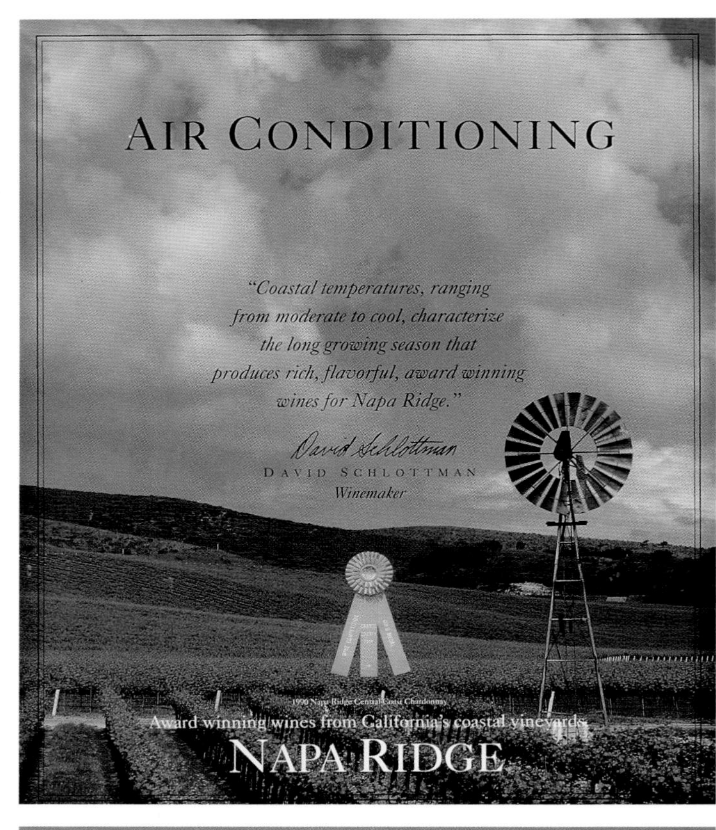

Nuance

Client
National Beverage Corporation

Firm
Michael Osborne Design

Art Director/Designer
Michael Osborne

Illustrator
Carolyn Vibbert

Cranberrie
Yummy & Rummie

Client
Chittenden Kitchens

Firm
SullivanPerkins

Art Director
Ron Sullivan

Designer
Paul Brouillette

Photographer
Gerry Kano

Bartlett Maine
Estate Winery

Client
Bartlett Maine
Estate Winery

Firm
Louise Fili Ltd.

Art Director/Designer
Louise Fili

Photographer
Austin Hughes

Bartlett Maine
Estate Winery

Client
Bartlett Maine
Estate Winery

Firm
Louise Fili Ltd.

Art Director/Designer
Louise Fili

Photographer
Austin Hughes

BARTLETT
Raspberry Wine
sweet
Alcohol 11.5% by Volume

BARTLETT
Wild Blueberry Wine
semi-dry
Alcohol 11.5% by Volume

Knapp Wine Labels

Client
Knapp Vineyards

Firms
Louise Fili Ltd.
Riley Associates

Art Director/Designer
Louise Fili

Illustrator
Philippe Weisbecker

Pasta La Bella

Client
American Italian
Pasta Co.

Firms
Pedersen Gesk

Designers
Kris Morgan,
Richard McGowan

Art Director
Kris Morgan

Illustrator
Dave Albers

Photographer
Courtesy of
Pedersen Gesk

Biscotti Di Roma

Client
Award Baking
International

Firms
Pedersen Gesk

Art Director
Sig Gesk

Designer
Tjody Overson-de-Vaal

Photographer
Courtesy of
Pedersen Gesk

Catamount

Client
Catamount Brewing Co.

Firm
Sullivan and
Brownell, Inc.

Designer
Thomas Brownell

Illustrator
Linda Gallo

Photographer
Jack Rowell

CATAMOUNT
PORTER
Fresh-All Natural
BREWED IN VERMONT
STRICTLY FRESH • FINEST INGREDIENTS
PRIDE OF THE NORTH COUNTRY
BREWED & BOTTLED BY CATAMOUNT BREWING COMPANY, WHITE RIVER JCT. VERMONT
12 FL. OZ. (355 ML)

Home Spun Values

Food is the ultimate symbol of comfort, for some it is as reassuring as a warm bath or cozy bed. The kitchen cabinet is home for products that are so much a part of our lives that their familiar mascots, the Quaker Oats' Man, the Campbell's Kids, Aunt Jemima, Uncle Ben, and even the Kool-Aid Pitcher are friends who satisfy appetites and soothe the soul. Certain foods so immediately define home—jam, jelly, cereal, peanut butter, tuna fish, and, oh yes, mayonnaise—that when served they trigger Pavlovian memories of the breakfast nook and dinner table.

Since the introduction of packaged goods, the problem for graphic designers has been to imbue the presentation of these foods with a kind of domestic warmth. Homespun food packaging must suggest the personal touch—the attention to quality that emanates from the bosom of the family; the same loving care from mom, grandma or aunt Sally, or so the myth goes. The fact is, since today's packaged goods contain products that have as much kinship to hearth and home as a hot dog has to steak, these packages are driven by mythologies that many of us have never really experienced. How many of us have time to devote to "loving" food preparation? These days the closest most people get to a personal exchange between product and consumption is the supermarket checkout clerk who indifferently scans the day's purchases.

In the absence of the neighborhood vendor or grocer, packaged goods that suggest homespun wholesomeness have a definite allure. By necessity consumers are forced to put their trust into packages that claim to embody the ideals that one hopes are true. Take for example, one of the most popular mainstream brands: with its idyllic renderings of rural scenes and horse-drawn carriages, Pepperidge Farm's baked goods project the halcyon days of pre-consumerist America. Who wouldn't select this over many similar generic products? The introduction of homespun conceits tells a story that pulls on the heartstrings as it appeals to the appetite. Yet once upon a time, homespun was considered too old-fashioned for the new era in food technology. In the late 1930s brands like Wonder Bread were given graphics that emphasized the sanitary and vitamin-enriched natures of products that came off the assembly line. The notion that science was modernity, and modernity was a virtue in everything from automobiles to bread held sway. Today there are few virtues afforded the assembly line, especially in food production, and both quaintness and nostalgia have returned.

This section is not, however, concerned with the tried-and-true packages that Americans have learned to trust, but rather, current solutions to the problem of injecting homespun values into relatively new brands. And of all the examples, two represent the extremes of the homespun spectrum. Sharon Werner's design for the Archer Farms milk carton complements the nutrition of mom's beverage of choice in a totally modern manner. The no-nonsense white carton clearly shows the product's name, contents, and fat percentage through bold sans serif typography. Since fat has become a national obsession in most households, the clear display of this information is reassuring, not just in a scientific way, but the way that mom might say it. Moreover, the subtle color scheme also telegraphs that this product is fresh, safe and healthy. On the other side of the spectrum, the Bees Knees honey package presents a nostalgic coziness. The quaint name is complemented by the hand-drawn bee and honeycomb pattern suggesting the all-natural origins of this product.

Today, homespun value added design runs the gamut from traditional, folksy graphics to modern reinterpretations. In all cases the goal is to appeal to the nurturing part of the consumers and to reassure them that what they buy is 100 percent good for the entire family.

Sweet Dark Chocolate

Client
Newman's Own Organics

Firms
Coulson Design
Newman's Own Organics

Art Director
Peter Meehan

Designer/Illustrator
Tauna Coulson

Production Artist
Jasper Lawrence

Photographer
Thomas Burke

Food Services of America Pomodori Tomatoes

Client
Food Services of America

Firm
Hornall Anderson Design Works

Art Director
Jack Anderson

Designers
Jack Anderson
Mary Hermes

Photographer
Tom McMackin

3oz. Chocolate Bar

Client
Newman's Own Organics

Firms
Coulson Design
Newman's Own Organics

Art Director
Peter Meehan

Designer/Illustrator
Tauna Coulson

Production Artist
Jasper Lawrence

Photographer
Thomas Burke

NEWMAN'S OWN™
ORGANICS
ALL PROFITS
After
Taxes
GO TO CHARITY
NET WT 3 OZ (85g
CHOCOLATE

NEWMAN'S OWN™
ORGANICS
ALL PROFITS
After
Taxes
GO TO CHARITY
CHOCOLATE

NEWMAN'S OWN™
ORGANICS
ALL PROFITS
After
Taxes
SWEET DARK CHOCOLATE

NEWMAN'S OWN™
ORGANICS
ALL PROFITS
After
Taxes
GO TO CHARITY
MILK CHOCOLATE
NET WT 3 OZ (85g)

Naturally Spaghetti Sauce

Client
Grand Palace Foods
International

Firm
Supon Design Group

Art Director
Supon Phornirunlit

Designer/Illustrator
Richard Lee Heffner

Photographer
Oi Jakrarat Veerasarn

Grafton Goodjam
Vinegar

Client
Grafton Goodjam

Firm
Louise Fili Ltd.

Art Director/Designer
Louise Fili

Photographer
Spiro/Bodi Productions

Red Wine Vinegar
Sweet BASIL
Peppers. Garlic
Peppercorns

32ounces

Vinegar Line II

Client
Grafton Goodjam

Firm
Grafton Goodjam

Art Director/Designer
Mary Schoener

VINAIGRE À
L' ÉCHALOTE
CHAMPAGNE VINEGAR
Shallots & Peppercorns
16 FL OZ (473 ml)
GRAFTON GOODJAM · GRAFTON VT 05146

Vinegar Line II

Design Firm
Grafton
Goodjam

Designer
Mary Schoener

White Wine Vinegar
Sage
Peppers. Shallots
Peppercorns

16 ounces

TARRAGON VINEGAR
APPLE CIDER VINEGAR·TARRAGON
25 OUNCES
GRAFTON GOODJAM • GRAFTON VERMONT 05146

APPLE CIDER VINEGAR
BASIL·LEMON BALM·GARLIC·TARRAGON
25 OUNCES
GRAFTON GOODJAM • GRAFTON VERMONT 05146

Vinegar Line II

Client/Firm
Grafton Goodjam

Designer/Illustrator
Mary Schoener

Jam

Client
Grafton Goodjam

Firm
Grafton Goodjam

Art Director
Mary Schoener

Designer
Martin Moskof

Apple Shrub

Client
Tait Farm Foods

Firm
Collins & Collins
Graphic Design

Art Director/Designer
Gretl Collins

Beth's Babies

Client
Beth's Fine Desserts, Inc.

Firm
Hilary Mosberg Illustration & Design

Art Director/ Designer
Hilary Mosberg

Photographer
Beth Setrakian

Clover Honey

Client/Firm
The Bee Knees by Ballard Apiaries

Photographer
Bob Ballard

Vermont Gold
Maple Syrup

Client
Chateau Nicholas

Firms
Zu Design
Chateau Nicholas

Art Director
Peter Vogel

Designer
Taro Masuda

Illustrator
Anthony Russo

Photographer
Chris Vaccaro

Lactantia PurFiltre Milk

Client
Ault Foods Limited

Firm
Tudhope Associates Inc.

Creative Director
Ian Tudhope

Designers
Cathy Russell
Creeshla Hewitt

Illustrator
Damian Glass

Lactantia All Natural Ice Cream

Client
Ault Foods Limited

Firm
Tudhope Associates Inc.

Creative Director
Ian Tudhope

Designer
Creeshla Hewitt

Illustrators
Gerrard Gauci
Sylvie Daigneault
Jim Stewart
Ricardo Stampotori
Raphael Montoliu

Lactantia All Natural
Frozen Yogurt

Client
Ault Foods Limited

Firm
Tudhope Associates Inc.

Creative Director
Ian Tudhope

Designers
Creeshla Hewitt
Ava Abbott

Illustrators
Lauri Lafrance
Vince McIndoe
James Marsh
Raphael Montoliu

Honey Acres Packaging & Collateral

Client
Honey Acres

Firm
Paula Fortney & Associates

Designer/Illustrator
David Deady

Photographer
Keith Meiser

BEEKEEPER'S BEST
HONEY
ACRES
HONEY ACRES
CLOVER HONEY
HONEY APRICOT
WILDFLOWER HONEY
HARVEST SAMPLER
BASSWOOD HONEY
NET WT 5.5 OZ (156

Archer Farms

Client
Target Stores

Firm
Werner Design
Werks, Inc.

Designers
Sharon Werner

Todd Bartz

Photographer
Darrell Eager

Vermont Milk

Client
Vermont Milk Producers

Firm
Sullivan and
Brownell, Inc.

Art Director
Kent Gardner

Designer
Thomas Brownell

Illustrator
Roger Beerworth

Photographer
Jack Rowell

Nostalgia *and* Vernacular

Collectable cookie tins, bottles, and packages command increasing prices from antique dealers, but the mass consumer is also intrigued by the graphics of the past. The surge in specially issued, limited edition packages (of millions) from Kellogg's, Hershey's, and other mainstream food companies showing how their venerable brands were originally packaged, has piqued the interest of consumers who have grown tired of the standard "new and improved" type of hype. Because many of these designs are more prosaic than the majority of contemporary food graphics, they appeal to the fashion conscious among a new generation of customers while pampering the nostalgic sides of the old ones. Underscoring a brand's rightful place in the history of American food production, these authentic, old packages (often adorned with the original logos) become symbols of quality.

Nostalgia has been a very successful marketing tool that derives, in part, from over-saturation of design conventions in an age where contemporary packaging has become wallpaper to the average consumer. By taking a particular brand back to its roots through either real or simulated nostalgic graphics, the consumer is given an alternative to the monotony on the shelves. At the same time, a particular brand is imbued with authority rooted in a heritage that demands it deserves the consumer's trust. But not all nostalgic packaging is drawn from real history. While the venerable companies dust off their prized antiques, younger ones invent their own histories by simulating old-fashioned graphics. A notable example is Crabtree & Evelyn, the English cosmetics, sundries, and food manufacturer which was founded in the early 1970s. Borrowing from nineteenth-century traditions, they introduced various new brands in quaint Victorian and Olde English packages, at once evoking a sense of national pride and a fairy-tale environment. The strategy worked, for not only did the products do extremely well in the specialty or boutique market, but Crabtree & Evelyn inspired competitors to follow the leader into the bowels of what is commonly called "retro."

As an offshoot of nostalgia, retro is the adaptation of distant and recent historical motifs placed in a contemporary context. The quintessential examples of retro today are 1940s and 1950s graphic elements, such as stock printer's illustrations (known as clichés), ornamental boarders, and novelty typefaces that approximate period styles as in camp or parody. Retro is a kind of movie set, fantasy sensibility that transports baby boomers (the primary audience for most food products today) back to those real and imagined epochs when life was more innocent—and when advertising graphics were more naive. The fact is that life was not any more innocent than it is today, but it is the nature of nostalgia to smooth out the rough edges and flatten out memory into a smooth, unblemished surface. Retro is an effective way to entertain and at the same time communicate a message without pushing too hard on the envelope of acceptability.

Retro is a transient state that rides on the nostalgic wave of the moment and is easily dismissed at the first sign of a new trend. What is nostalgic today will be passé tomorrow. For retro to be effective it must sustain a sense of camp; once it becomes a common conceit, a permanent fixture on a package, it is no longer retro. Therefore, retro is always in a state of reevaluation. Likewise, vernacular is constantly in flux. Vernacular is a common graphic vocabulary based on a general acceptance of a particular visual language or style. Vernacular can be retro, such as the vernacular of the 1950s that was rooted in specific type and graphic images unique to that period, but it can also be contemporary, as in generic styles applied to current products. The common or standard methods of designing mass market items, such as soda, candy, and cereal constitute a vernacular of contemporary packaging. Graphic designers either submit to the vernacular and follow the pack, or manipulate these stereotypical approaches to transcend the commonplace. The most imaginative designers satirize or parody the vernacular while at the same time present the product to the appropriate target audience.

Nostalgia and vernacular appeal to the consumer's fantasies of a near or distant past be they fictitious or real. The challenge is not to misuse the tools so that a product is out of the past, but rather becomes a commentary on the past in a modern setting.

Il Fornaio Coffee, Pasta, Baked Goods & Collateral

Client
Il Fornaio

Firm
Michael Mabry Design Inc.

Designer/Illustrator
Michael Mabry

Photographer
Michael Lamotte

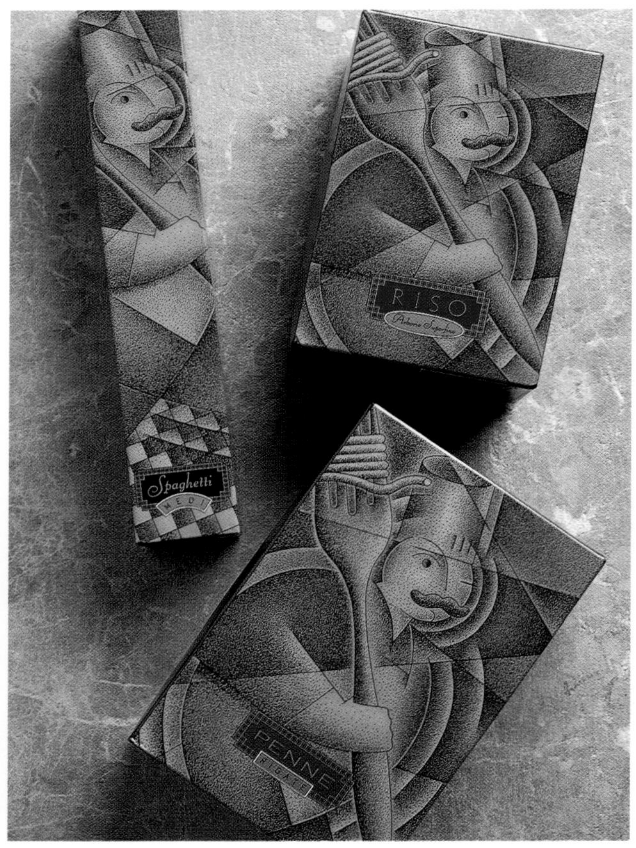

Il Fornaio

Cyrk Pure &
Natural Gourmet Ice
& Ice Cream

Client
Cyrk Fine Foods

Firm
The Pushpin Group, Inc.

Art Director
Seymour Chwast

Designer
Roxanne Slimak

Photographer
Ed Spiro

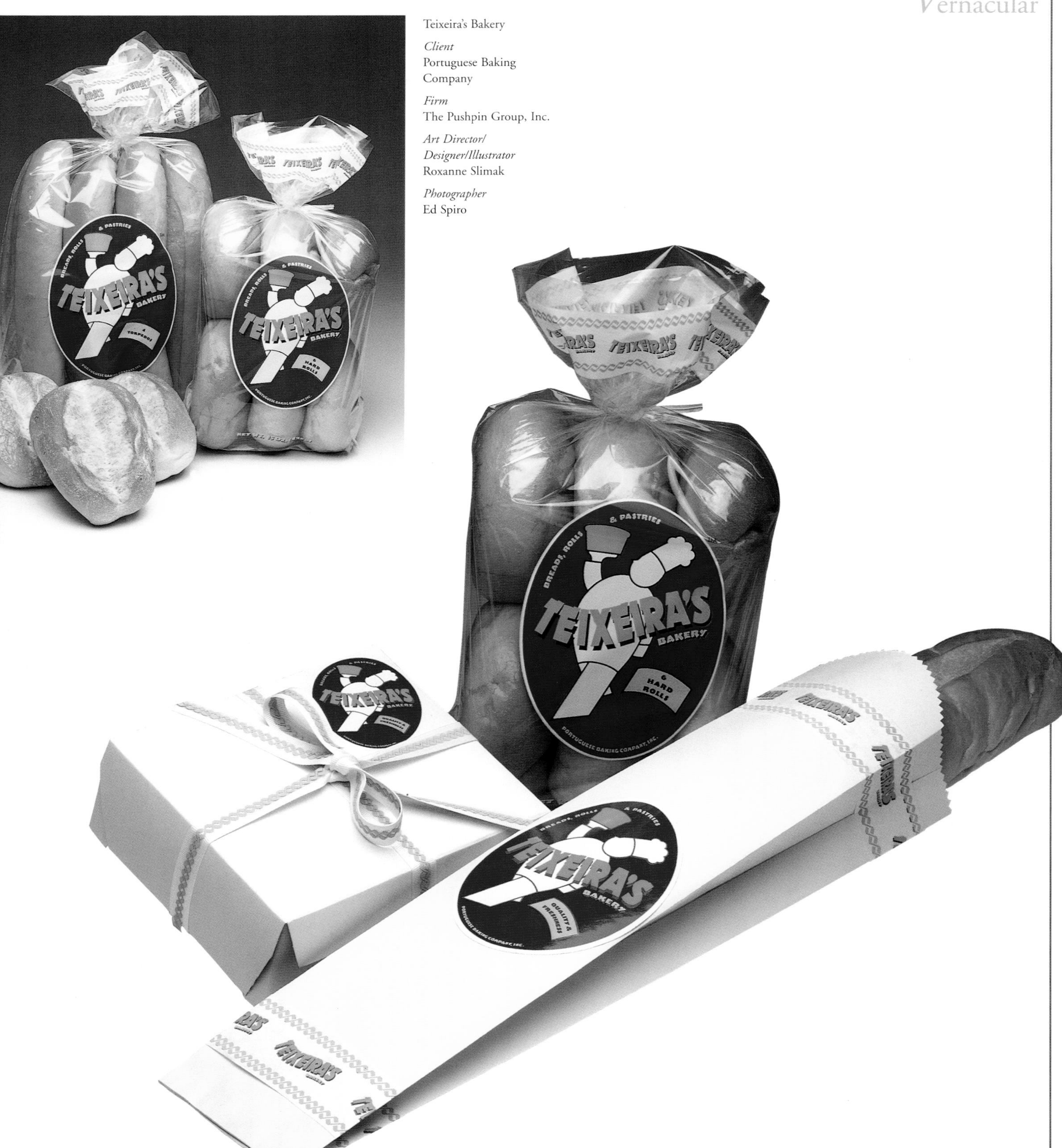

Teixeira's Bakery

Client
Portuguese Baking Company

Firm
The Pushpin Group, Inc.

Art Director/ Designer/Illustrator
Roxanne Slimak

Photographer
Ed Spiro

Buckaroo Gourmet Foods Packaging

Client
Best of the West, Inc.

Firm
Thornton and Associates

Art Director/Designer/ Illustrator
Gary Thornton

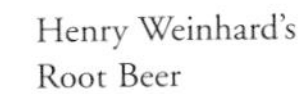

Henry Weinhard's
Root Beer

Client
G. Heileman
Brewing Company

Firm
Primo Angeli Inc.

Art Directors
Carlo Pagoda
Primo Angeli
Brody Hartman

Designers
Mark Jones
Brody Hartman

Illustrator
Rick Gonella

Typographer
Sherry Bringham

Production Manager
Eric Kubly

Photographer
June Fouche

Arizona Iced Teas

Client
Arizona Iced Teas

Firms
Jean Pettine Graphic
Design
Arizona Iced Teas

Art Director
Jean Pettine

Designers
John Ferolito
Don Vultaggio

Photographer
Bill Streicher

Busha Browne's
Jamaican Produce

Client
Busha Browne's
Company Ltd

Firm
Michael Thierens Design

Designer
Michael Thierens

Illustrator
Ian Beck

Photographer
Laurie Collard

BUSHA BROWNE'S
Fine JAMAICAN Produce
The BUSHA BROWNE Story
LIME PEPPER JELLY
SPICY & HOT PEPPER SHERRY
SPICY TOMATO LOVE-APPLE SAUCE
original PLANTERS spicy piquant sauce
NET WT 6⅛ fl oz 184 ml
ORIGINAL GUAVA JELLY TWICE BOILED
original BANANA chutney
NET 12 oz 340 g WT
SPICY FRUIT chutney
ORIGINAL GUAVA JELLY
TWICE BOILED
NET WT
12 oz 340 g
BURNED ORANGE MARMALADE
BANANA chutney
TWICE BOILED GUAVA JELLY
BURNED ORANGE MARMALADE

Bread Mix

Client
Trillium Health Products

Firms
Studio Bolo
Trillium Health Products

Art Director
Cindy Winemiller

Designer
James Forkner

Illustrator
Carolyn Vibbert

Copywriter
Kit Hutchin

Photographer
Tyler Boley

Rice

Client
Trillium Health Products

Firms
Studio Bolo
Trillium Health Products

Art Director
Cindy Winemiller

Designer
James Forkner

Illustrator
Carolyn Vibbert

Copywriter
Kit Hutchin

Photographer
Tyler Boley

B&G Beaujolais Nouveau

Client
Seagram Classics Wine Co.

Firm
Halleck Design Group

Designers
Ross Halleck
Paul Morales

Illustrator
Sue Rother

Red Sky Ale

Client
St. Stan's Brewing Company

Firm
Michael Osborne Design

Art Director
Michael Osborne

Designer/Illustrator
Chris Lehmann

Coors Special Lager

Client
Coors Brewing Company

Firm
WBK Design

Art Director
Tony Hyun

Designer
Doug Faulkner

Photographer
OMS Photography

Teeccino (Herbal Espresso) Packaging

Client
Teeccino Café, Inc.

Firm
Gina Amador Design

Art Director/Designer
Gina Amador

Illustrator
Jennifer Ewing

Cha Cha Chips
& Salsa

Client
Ham I Am

Firm
SullivanPerkins

Art Director/ Designer/Illustrator
Linda Helton

Photographer
Gerry Kano

Classical Dress

Food packages come in all shapes and sizes as well as various looks and styles. The mannerism called Classical Dress represents the elegant side of the spectrum, a sensibility rooted in the traditions of fine typography and simple geometry. While this comparatively subtle approach is only appropriate for certain kinds of upscale food and drink, it has nevertheless become a very popular way of clothing a product, particularly when the manufacturer wants to present an aura of exclusivity that commands a premium price.

In recent years Classical Dress has added value to some otherwise common products. Premium breakfast cereal, for example, the so-called healthy variety of reduced fat grains and berries has been distinguished from the sugar-coated, calorie-rich genre targeted at children by replacing goofy mascots and cartoon typefaces with simple still life photographs and exquisite Garamond and Bodoni typefaces. Among the gaudier packages, these typographic respites are welcome islands of clarity on the cluttered supermarket shelf. Likewise mass-produced teas, such as Lipton with its eye-catching fire-engine red and sunburst yellow package—once the standard against which kindred products were swathed—are today given the classical look befitting the old English tea drinking tradition. Even some generic and supermarket house brands have been designed with an eye towards subtlety. Amidst aisles of screaming packages the most sedate usually speak the loudest.

Such is the case with that trendiest of all contemporary drinks, bottled water, which today is the most common example of contemporary Classical Dress. Water is one of the most difficult products to distinguish in a market currently flooded with scores of different brands. Here is a genre where packaging and advertising separates the weak from the strong, the successes from the failures. This is also a genre where the only attribute that truly distinguishes one brand from another is how the public perceives the particular brand—which determines how marketers select the special niche for success. While economical waters come in bulky, inelegant plastic containers, Classical Dress is used to transform the common mixture of two parts hydrogen and one part oxygen into a virtual champagne.

How does water become something that fulfills one's deepest desire? Until recently, one would turn on the tap, fill up a glass and add a few ice cubes. Today, with the large selection of bottled waters available in any grocery or delicatessen, the idea of drinking water from the kitchen sink seems, well, somewhat plebeian. Such an array of natural and flavored waters begs the question why anyone would drink from the rusty old tap, anyway. Today, the bottled water industry commands such a large percentage of the non-alcoholic beverage market that packaging is the most significant factor in winning brand loyalty. Elegance is the key for those waters that appeal to the consumerist impulse which has more to do with status than health. Like a fine wine, an elegant bottle of water compliments any dining experience.

Among these eye-catching waters Tynant is the most seductive. A mesmerizing cobalt blue bottle projects an aura of mystery. While the bottle echoes the classic Perrier shape, and so trades on the equity of the leading brand, its unique deep blue hue gives Tynant the quality of nobility that its competitor does not have. The premium price charged for Tynant is not justified because the water is tastier than others, but that bottle projects such majesty that it demands greater sacrifice. The image has been so successfully orchestrated that empty Tynant bottles have become virtual objet d'art in homes and restaurants. For a container to be both artful and functional is a significant achievement, especially when the goal is mass consumption. This is why Adonis, designed by the Supon Design Group, is such a remarkable package. Targeted at the athlete, the designer employs a classical symbol of the ideal physical specimen. Adonis may not be marketed as a table water, but its elegant composition suggests that this is not only the perfect drink after a perspiring workout, but is a perfect complement after a hard day at the health club.

Whether the product is water, cereal, tea, ice cream, or candy, Classical Dress is a means of both asserting authority and maturity. From the coats of arms, borrowed from the past which adorn many classical looking products, to the expert contemporary application of fine typography this method elevates the product into an object of desire.

Pane di Paolo
Ciabatta & Piatto

Client
Broadmoor Baker

Firm
Hornall Anderson
Design Works

Art Director
Jack Anderson

Designers
Jack Anderson
Mary Hermes
Leo Raymundo

Illustrator
Larry Jost

Photographer
Tom McMackin

Napa Valley Oils

Client
Napa Valley Kitchens

Firm
Colonna Farrell Design

Art Directors
Ralph Colonna
Cynthia Sterling

Designer
Cynthia Sterling

Illustrator
Mike Gray

Photographer
©Photo by Teri Sandison

Big Town Picnic Wines

Client
Little City

Firm
Bruce Yelaska Design

Art Director/
Designer/Illustrator
Bruce Yelaska

Photographer
Tom McCarthy

Candinas Chocolates

Client
Candinas Chocolatier

Firm
Planet Design Compar

Designer
Kevin Y. Wade

Photographer
Mark Salisbury

CANDINAS CHOCOLATES
CANDINAS

Equatorial Provisions

Client
Chocolate Moon

Firm
Creative Forces

Designer
Clare Ahern

Zima

Client
Foote, Cone & Belding for
Adolph Coors Brewing Co.

Firms
Primo Angeli Inc.
Libby Perszyk Kathman
Adolph Coors Brewing Co.
Foote, Cone & Belding

Art Directors
Primo Angeli
Carlo Pagoda
Ray Perszyk
Howard McIlvain
Pam Moorehead
George Chadwick

Designers
Carlo Pagoda
Ed Cristman
Vicki Cero
Bob Johnson
Jim Gabel
Mary Jo Betz
Bradd Bush
Liz Grubow
Andy Scott
Rowland Heming

Àsanté

Client
National Beverage Corporation

Firm
Michael Osborne Design

Art Director/Designer
Michael Osborne

TAZO
Simply Red
TAZO
The Reincarnation of Tea
Passion Potion
[tea & juice]
The exotic taste of
passionfruit mixed with refreshing
microbrewed green tea.
16
fluid ounces
(475ml)
TAZO
The Reincarnation of Tea
Basic Black
A microbrewed
classic black tea with a light,
natural sweetness.
16
fluid ounces
(475ml)
TAZO

TAZO
Zen
TAZO
Calm
[caffeine free]
TAZO
Wild Sweet Orange
[caffeine free]
TAZO
Spice
[caffeine free]
TAZO
Passion
[caffeine free]
TAZO
Awake
TAZO
Earl Grey
TAZO
Refresh
[caffeine free]

TAZO
of TEA
Calm
TAZO

Tazo Tea

Client
Steve Smith

Firm
Sandstrom Design

Art Director
Steve Sandstrom

Designers
Steve Sandstrom
Janée Warren

Copywriter
Steve Sandoz

Photographer
Mark Hooper

Fife Wine

Client
Fife Vineyards

Firm
Colonna Farrell Design

Art Director
Cynthia Sterling

Designer
Amy Linn

Photographer
Richard Kasmier

Schwartzkopf
Commemorative Brandy

Client
Jepson Vineyards Ltd.

Firm
Glenn Martinez
and Associates

Art Director/Designer
Glenn Martinez

Photographer
Don Silverek

DVX Mumm Cuvée
Napa

Client
The Seagram
Classics Wine Co.

Firm
Michael Osborne Design

Art Director
Michael Osborne

Designer
Kristen Clark

Illustrator
Nicole Miller

Ty Nant Spring Water

Client
JML Importing Co., Inc.

Photographer
Ty Nant

Fassati Wine

Client
Marchesi Fassati Di Balzola

Firm
Vignelli Associates

Art Director/Designer
Massimo Vignelli

Photographer
Luca Vignelli

Hallmark Crown

Client/Firm
Hallmark Cards, Inc.

Art Directors
Gary Schenck
Jim Ramirez

Designer
Jim Ramirez

Photographer
Anne E. Blair

Soléo

Client
Sutter Home

Firm
Michael Osborne Design

Art Director/ Designer/Typographer
Michael Osborne

Paradise Rice Wine

Client
Grand Palace
Foods International

Firm
Supon Design Group

Creative/Art Director
Supon Phornirunlit

Designer
Andrew Berman

Photographer
Oi Jakrarat Veerasarn

Pororoca

Client
Plaza Design

Firm
Sussman/Prejza &
Co., Inc.

Architect
Fernando Vazquez

Designer
Deborah Sussman

Photographer
Annette del Zoppo
Productions

Picholine Olives

Client
Picholine

Firm
Louise Fili Ltd.

Art Director/Designer
Louise Fili

Illustrator
Anthony Russo

Photographer
Ed Spiro

The King's Cupboard

Client
West Fork Creations, Inc.

Art Director/Illustrator
Jerry Krauser

Designers
Jerry Krauser
Victoria Deiro

The Dip Kit

Client
Ham I Am

Firm
SullivanPerkins

Art Director
Ron Sullivan

Designer
Melissa Witt

Photographer
Gerry Kano

Harden & Huyse Chocolates

Client
Harden & Huyse

Firm
Primo Angeli Inc.

Art Directors
Primo Angeli
Carlo Pagoda

Designers
Philippe Becker

Production Manager
Eric Kubly

Photographer
June Fouche

"J" Jordan Sparkling
Wine and Gift Box

Client
Jordan Sparkling
Wine Company

Firm
Colonna Farrell Design

Art Director
Ralph Colonna

Designers
Ralph Colonna
Peggy Koch

Photographer
David Bishop/
San Francisco

Michael Osborne Design
Olive Oil

Client/Firm
Michael Osborne Design

Art Director/Designer
Michael Osborne

Rare Hawaiian White Honey

Client
Volcano Island Honey Co.

Firm
Volcano Island Honey Co.

Illustrator
Deitrich Varez

Domaine Chandon Brandy

Client
Domaine Chandon

Firm
Colonna Farrell Design

Art Director
Ralph Colonna

Designer
Cynthia Sterling

Calligrapher
Jeanne Greco

Photographer
Richard Kasmier

Consorzio Chiara

Client
Napa Valley
Kitchens

Firm
Michael Mabry
Design Inc.

Designer/Illustrator
Michael Mabry

Photographer
Michael Lamotte

Directory

Allegro Coffee Company
1930 Central Avenue
Boulder, CO 80301
303-444-4844
Fax: 303-449-5259

Allen & Cowley
Urban Trading Co.
4221 East Raymond #100
Phoenix, AZ 85040
602-437-1587
Fax: 602-437-3056
Lee Allen

Amazing Grazing
3001 Rockmart Road
Rome Road, GA 30161
706-802-0099
Sara Carpenter

American Italian Pasta Co.
1000 Italian Way
Excelsior Springs, MO 64024
816-630-6400

Anderson Design
30 North 1st Street
Minneapolis, MN 55401
612-339-5181
Charles Anderson

Arizona Beverages
5 Dakota Drive, Suite 305
Lake Success, NY 11042
513-357-4750
Fax: 513-357-4754

Atlantis Coastal Foods
Bubba Brand Coffee
708 King Street
Charleston, SC 29403
803-853-9444
Fax: 803-853-8463
Mike Zemke

Ault Foods Limited
405 The West Mall
Etobicoke, Ontario
M9C 511
Canada
416-620-3543
Fax: 416-620-3600
Kelly Gillespie

Award Baking International
1101 Stinson Boulevard
Minneapolis, MN 55401
612-331-3523
Fax: 612-331-1685
Marie Kennedy

Ballard Apiaries
Bees Knees
Box 324
Debolt, Alberta
T0H 1B0
Canada
403-957-2533
Fax: 403-957-2933
James Ballard, Vicky Ballard

Bartlett Maine Estate Winery
RR #1 Box 598
Gouldsboro, ME 04607
207-546-2408
Fax: 207-546-2554

Beekeeper's Best
Honey Acres
Ashippun, WI 53003
414-474-4411
800-558-7745
Eugene Brueggeman

Best of the West, Inc.
P.O. Box 160274
Austin, TX 78716-0274
512-327-3741
Gerry

Beth's Fine
Desserts & Company
1201 S Anderson Drive
San Raphael, CA 94901
415-485-1406
Fax: 415-457-6856
Beth Setrakian

Bette's Diner Products
4240 Hollis Street
Emeryville, CA 94608
510-601-6980
Bette Kroening

Bishop Photography
610 22nd Street
San Francisco, CA
415-558-9532
Fax: 415-626-7643
David Bishop

Bone Dry Beer Microbrewery
2255 Bancroft Avenue
Los Angeles, CA 90039
213-668-1055
Fax: 213-668-2470
Nancy Loose

Brand Equity International,
A Selame Company
2330 Washington Street
Newton, MA 02162
617-969-3150
Fax: 617-969-1944
Elinor Selame

Britches of Georgetown
544 Herdon Parkway
Herdon, VA 22070
703-471-7900
Susie Egan

Broadmoor Baker
Skinner Building
735 Fifth Avenue
Seattle, WA 98101
206-624-0000
Paul Suzman

Bruce Yelaska Design
1546 Grant Avenue
San Francisco, CA 94133
415-392-0717
Fax: 415-397-1174

Michael W. Buckley
503 Underhill Avenue
Yorktown Heights, NY 10598
914-962-5314
Fax: 914-962-5375

Buffalo Bill's Brewery
1082 B Street
Hayward, CA 94541
612-722-8050
Bill Owens

Busha Browne's Company Ltd.
P.O. Box 386, Newport East
Kingston, Jamaica
West Indies
809-923-8911-8
Fax: 809-923-9132
Winston Stona

Cafe Marimba
2317 Chestnut Street
San Francisco, CA 94123
415-982-7336
Fax: 415-398-6426

Cafe Society Coffee Company
2910 N Hall Street
Dallas, TX 75204
214-922-8888
Fax: 214-922-0280
Laurie Sandefer

Candinas Chocolatier
2435 Highway PD
Verona, WI 53593
608-845-1545
Fax: 608-845-1555
Markus Candinas

Capons Rotisserie Chicken
605 15th Avenue East
Seattle, WA 98102
206-282-0398

Carolina Swamp Stuff
4050 Arendell Street
Morehead City, NC 28557
919-247-9267
Fax: 919-247-9560
Mel Hughes

Catamount Brewing Co.
58 South Main Street
White River Junction, VT 05001
802-296-2248
Fax: 802-296-2420

Chateau Nicholas
RD4 Box 867
Brattleboro, VT 05301
802-254-5529
Fax: 802-254-5649
Peter Vogel

Chittenden Kitchens
Mountain Top Road
HC 32 Box 47
Chittenden, VT 05737
802-483-2979
802-483-6161
Fax: 802-483-6599
Kathy Hall

Chocolate Moon
146 Church Street
Asheville, NC 28801
704-253-6060
Fax: 704-253-1020
Robert Maddix

Chris May Design
200 East Ferry Road
Isle of Dogs
London
England
0171 515 0507

Classical Foods
45 Pondfield Road West #4F
Bronxville, NY 10708
914-779-6723

Cloud Nine Inc.
300 Observer Highway
Hoboken, NJ 07030
201-216-0382
Fax: 201-216-0383
Josh Taylor

Coburg Dairy, Inc.
5000 La Cross Road
North Charleston, SC 29406
803-554-4870
Fax: 803-745-5502

The Coca-Cola Company
One Coca-Cola Plaza
Atlanta, GA 30313
214-357-1781

The Coffee Plantation
1235 S 48th Street
Tempe, AZ 85281
602-966-9442
Fax: 602-966-9274
Joe Johnston

Colonna Farrell Design
1299 Main Street
Saint Helena, CA 94574
707-963-5865
707-963-5756
Ralph Colonna
Cynthia Macguire, Chris Mathes
Baldwin, Mike Gray,
Peggy Koch, Cynthia Sterling,
Amy Linn

Coors Brewing Company
311 North 10th Street
Golden, CO 80401
303-279-6565
Greg Head

Coulson Design
108 Locust Street, Suite A
Santa, Cruz, CA 95060
408-458-1166
Fax: 408-458-0336
Tana Coulson

Creative Forces
287 Josephine Street NE
Atlanta, GA 30307
404-524-6780
Fax: 404-524-7076
Clare Ahern

Crystal Geyser
Washington Street
Calistoga, CA 94515
Peter Gordon

Culinary Alchemy Inc.
P.O. Box 393
Palo Alto, CA 94302
415-367-1455
Fax: 415- 367-0223
Kathleen O'Rourke,
Elizabeth R. Connolly

Cyrk Fine Foods
162 Route 59
Monsey, NY 10952
914-356-3006
Bruce Serkez

Daka International
1 Corporate Place
55 Ferncroft Road
Danvers, MA 01923
508-774-9115

Debenport Photography
2412 Converse Street
Dallas, TX 75207
214-631-7606
Fax: 214-951-0424
Robb Debenport

Denzer's Food Products
P.O. Box 5632
Baltimore, MD 21210
410-889-1500
800-224-2811
Fax: 410-235-7032
Jacob W. Slagle

Domaine Chandon
California Drive
Yountville, CA 94599
707-944-8844
Fax: 707-944-1123

Doug Petty Studio
Skidmore Fountain Building
Portland, OR 97204
503-225-1062

Todd Eberle
54 West 21st Street
New York, NY 10010
Mark Ebsen

Egg Design Partners
790 Centre Street
Boston, MA 02130
617-522-7558
Fax: 617-522-7549

El Paso Chili Company
909 Texas Avenue
El Paso, TX 79901
915-544-3434
Fax: 915-544-7552
Park Kerr

Empress
International Limited
10 Harbor Park Drive
Port Washington, NY 11050
516-621-5900

Equitorial Provisions
146 Church Street
Asheville, NC 28801
800-723-1236
704-253-6060
Fax: 704-253-1020
Robert Maddix

Estudio Ray
2320 N 58th Street
Scottsdale, AZ 85257
602-945-1998
Fax: 945-1299
Christine Ray, Joe Ray,
Leslie Link, Frank Ybarra

Chip Fisher
27 East 92nd Street
New York, NY 10128

Food Services of America
4025 Delridge Way SW
Seattle, WA 98106
206-933-5000
Roger Toomey

Foote, Cone & Belding for
Adolph Coors Brewing Co.
1255 Battery Street
San Francisco, CA 94111
George Chadwick (FCB)

Fortune Design Studio
16932 NE 141st Place
Woodinville, WA 98072
206-483-5953
Fax: 206-483-9366
Lei Lani Fortune, John Fortune,
Lance Lindell

Found Graphic Design
3401 Crossland Avenue
Baltimore, MD 21213
410-235-7356

Foxhead Farms
108 Sweetwater Lane
Barnwell, SC 29812
803-259-5391
Drew Weeks

G. Heileman Brewing Company
Primo Angeli Inc.
590 Folsom Street
San Francisco, CA 94105
415-974-6100
Fax: 415-974-5476

Gebhart Productions, Inc.
8443 Warner Drive
Culver City, CA 90232
310-280-0650
Chuck Gebhart

Gil Shuler Graphic Design
231 King Street
Charleston, SC 29401
803-782-5770
Fax: 803-577-9691
Gil Shuler, Jay Parker,
Emily Bidwell, Steve Lepre

Gina Amador Design
303 California Avenue
San Rafael, CA 94901
415-456-6605
Gina Amador

Glenn Martinez and Associates
610 Davis Street
Santa Rosa, CA 95401
707-526-3198
Fax: 707-526-3213

Grafik Communications
1199 N. Fairfax Street
Suite 700
Alexandria, VA 22314
703-683-4686
Fax: 703-683-3740
Judy Kirpich, Michael Shea,
Susan English, Melanie Bass

Grafton Goodjam
RD 3
Grafton, VT 05146
802-843-2100
Fax: 802-843-2589
Mary Schoener

Haley Johnson Design Co.
3107 East 42nd Street
Minneapolis, MN 55406
612-722-8050
Fax: 612-722-5989

Halleck Design Group
470 Pamona Street
Palo Alto, CA 94301
415-325-0707
Fax: 415-325-0738

Hallmark Cards, Inc.
2501 McGee Box 419580
Kansas City, MO 64141
Anne E. Blair

Ham I Am
1303 Columbia Drive
Suite 201
Richardson, TX 75081

Harden & Huyse
Primo Angeli Inc.
590 Folsom Street
San Francisco, CA 94105
415-974-6100
Fax: 415-974-5476
Jean Galeazzi

Il Fornaio
1000 Sansome Street
San Francisco, CA 94111
415-986-1505
Fax: 415-986-2879
Hilary Wolf

James Collins Graphics
557 Glenn Road
State College, PA 16803
814-234-2916
Gretl Collins

James Dorn of
PAC National Inc.
17735 NE 65th Street
Redmond, WA 98052
206-869-6869

Jean Pettine Graphic Design
202 High Street
Mapleshade, NJ 08052
609-667-4888

Jepson Vineyards Ltd.
10400 South Highway
Ukiah, CA 95482
707-468-8936
Fred Bellows

JML Importing Co. Inc.
34197 Pacific Coast
Highway #100
Dana Point, CA 92629
714-248-8208
Fax: 714-248-7793
James Lloyd, Jepson Brandy

Joel Levin Photography
1530 15th Avenue
Seattle, WA 98122
206-323-7372

Jordan Vineyards
150 North Street
P.O. Box 1919
Healdsburg, CA 95448
707-431-5200
Fax: 707-431-5207

KAZ Photography
215 Adobe Canyon Road
Kenwood, CA 95452
707-833-2536
Fax: 707-833-1241
Richard Kasmier

Keeven Photography
2838 Salena Street
St. Louis MO 63118
314-773-3587
Fax: 314-771-4808
Marty Keeven

Kimberly Baer
Design Associates
620 Hampton Drive
Venice, CA 90219
310-399-3295

Laster & Miller
2212 Yandell Street
El Paso, TX 79903
915-533-9800
Fax: 915-544-4069

Lipson Alport
Glass & Associates
666 Dundee Road, Suite 103
Northbrook, IL 60062
847-291-0500
Fax: 847-291-0516
Sam J. Cuilla, Tracy Bacilek,
Carol Davis, Amy Russell

Little City
673 Union Street
San Francisco, CA 94133
415-434-2900
Herb Beckman

Lombardi Dry Ice Selzer
330 Willow Street
New Haven, CT 06511
800-770-3302
Frank Vollero

Louise Fili Limited
71 Fifth Avenue
New York, NY 10003
212-989-9153
Fax: 212-989-1453
Louise Fili

Love Packaging Group
410 E 37th Street N
Wichita, KS 67201
316-832-3369
Fax: 316-832-3293
Tracy Holdeman

M.A. O'Halloran
258 Collins Street #3
San Francisco, CA 94118
415-387-8398

Madame Sophie's
370 State Street
North Haven, CT 06473
203-287-0227
Fax: 203-785-8758
Karen E. Burgess

Mangia Chicago Stuffed Pizza
2401 Lake Austin Boulevard
Austin, TX 78703
512-478-6600

Marchesi Fassati Di Balzola
via Pattari 6
Milan 20122
Italy
Leonardo Fassati

Margo Chase Design
2255 Bancroft Avenue
Los Angeles, CA 90039
213-668-1055
Fax: 213-668-2470
Margo Chase, Wendy Ferris, Anne Burdick

Mario Parnell Photography
301 8th Street #202
San Francisco, CA 94103
415-553-8555
Fax: 415-431-5670

Tom McCarthy
Tom McCarthy Photography
480 Gate 5 Road #259B
Sausalito, CA 94965
415-986-0525

McCleary & Company
3714 34th Avenue SW
Seattle, WA 98126
206-938-3770
Fax: 206-932-1876
Glenda McCleary

Merchant du Vin Corporation
140 Lakeside Avenue
Suite 300
Seattle, WA 98122-6538
206-720-2209

Michael Fioritto Photography
3728 East Emerald Street
Mesa, AZ 85206
602-924-6304

Michael Lamotte
Studios Inc.
424 Treat Street
San Francisco, CA 94110
415-431-1443
Fax: 415-431-6044

Michael Mabry Design
212 Sutter Street
San Francisco, CA 94108
415-982-7336
Fax: 415-398-6426

Michael Osborne Design
444 DeHaro Street
San Francisco, CA 94107
415-255-0125
Fax: 415-255-1312
Michael Osborne, Chris Lehman, Tom Kamegai, Kristen Clark

Michael Thierens Design
Dee Road
Richmond, Surrey
England
181-332-6788
181-332-7340

George Mundell, III
621 Linden Road
Bellingham, WA 98225
360-733-0293
George Mundelli

Nantucket Nectars
100 Holton Street
Boston, MA 02135
617-789-4300
Fax: 617-789-5491
Charlie Conn

Nantucket
Offshore Seasonings
P.O. Box 1437
Nantucket, MA 02554
508-228-9292
Fax: 508-325-6265
Nigel Dyche

Napa Valley Kitchens
1236 Spring Street
Saint Helena, CA 94574
707-967-1107
Fax: 707-967-1117

Napa Valley Pantry
P.O. Box 50
Oakville, CA 94562
707-967-9176
Kelly Wheeler

National Beverage Corporation
One North University Drive
Fort Lauderdale, FL 33324
305-581-0922
Fax: 305-473-4710

Newman's Own Organics
The Second Generation
P.O. Box 2098
Aptos, CA 95001
408-685-2866
Fax: 408-685-2261
Peter Meehan

Office of Michael Manwaring
111 Crescent Road
San Anselmo, CA 94960
415-458-8100
Fax: 415-458-8150

Olde Colony Bakery
280 King Street
Charleston, SC 29401
803-722-2147
Sheila Rix

Palais D'Amour Honey
4729 31st Avenue South
Minneapolis, MN 55406
612-722-8623
Geyy D'Amour

Paula Fortney and Associates
1800 S Pierio Street
Chicago, IL 60608
312-829-6600
Fax: 312-829-7770
Paula Fortney

Peaberry Coffee Ltd.
4785 Elati Street
Denver, CO 80216
303-292-9324
Fax: 303-292-5179

Pedersen Gesk
105 5th Avenue
Minneapolis, MN 55401
612-332-1331
Fax: 612-332-7231

Pentagram Design
212 Fifth Avenue
New York, NY 10010
212-683-7000
Fax: 212-532-0181
Paula Scher, Ron Louie, Douglas Smith

The Perrier Group of America
777 West Putnam Avenue
Greenwich, CT 06836
203-531-4100
Andrew Carter

Picholine
35 West 64th Street
New York, NY 10023
212-724-8585
Fax: 212-875-8979

Piedmont Label Company
311 West Depot Street
Bedford, VA 24523
703-586-2311

Pirtle Design
56 West 22nd Street, 8th floor
New York, NY 10010
212-647-9870
Leslie Pirtle

Planet Design Company
605 Williamson Street
Madison, WI 53703
608-256-0000
Fax: 608-256-1975
Kevin Y. Wade

Primo Angeli Inc.
590 Folsom Street
San Francisco, CA 94105
415-974-6100
Fax: 415-974-5476
Primo Angeli, Philippe Becker, Carlo Pagoda, Rick Gonella, Sherry Bringham, Mark Jones, Brody Hartman, Ed Cristman, Vicki Cero

The Pushpin Group, Inc.
215 Park Avenue South
New York, NY 10003
212-674-8080
Fax: 212-674-8601
Seymour Chwast
Roxanne Slimak

RBMM—Dallas
7007 Twin Hills, Suite 200
Dallas, TX 75231-5184
214-987-6510
Fax: 214-987-3662

RDO Specialty Foods
472 Tehama Street
San Francisco, CA 94103
415-777-2578
Fax: 415-777-1750
Mary Anne O'Halloran

Real Restaurants
180 Harbor Drive, Suite 100
Sausalito, CA 94965
415-331-9101
Fax: 415-331-9022

The Republic of Tea
Eight Digital Drive, Suite 100
Novato, CA 94949
415-382-3400
Fax: 382-3401
Gina Amador, Shannon Connelly, Nancy Bauch, Vic Zauderer, Patricia Ziegler, Faye Rosenzweig

Rick's Kitchen
P.O. Box 1702
Highway 1075
Cashiers, NC 28717
704-743-2272
Rick Brooks

Robert Biale Vineyards
2040 Brown Street
Napa, CA 94559
707-257-7555

Rock Island Studios
337 N. Rock Island Studios
Wichita, KS 67202
316-263-8151
Don Siedhoff

Rosenworld
45 Lispenard Street #7E
New York, NY 10013
212-966-6896
Laurie Rosenwald

St. Stan's Brewing Company
821 L Street
Modesto, CA 95354
209-524-2337
Fax: 209-524-4827

San Anselmo's Biscotti
158 Paul Drive
San Raphael, CA 94903
800-229-1249
415-492-1220
Fax: 415-492-1282
Jane Cloth-Richman

Teri Sandison
Teri Sandison Photography
387 La Fata Street
St. Helena, CA 94562
707-963-3319
Fax: 707-963-1482

Steve Sandstrom
Sandstrom Design
808 SW Third Avenue
Portland, OR 97204
503-248-9466
Fax: 503-227-5035

The Seagram Classics
Wine Company
P.O. Drawer 500
8445 Silverado Trail
Rutherford, CA 94573
707-942-3406
Fax: 707-942-3469

Seaperfect, Inc.
2107 Folly Road
Charleston, SC 29412
803-762-5390
Tom Royall, Sally Burry

Sharon Till Associates
55 Francisco Street, Suite 450
San Francisco, CA 94133
415-421-6664
Fax: 415-421-8065

Shoebox Greetings
2400 Pershing Road, Suite 500
Kansas City, MO 64108
816-274-7629
Fax: 816-274-5818
John Wagner, Julie McFarland, Karen Brunke, Meg Cundiff

Sidney Cooper Photography
1427 East Fourth Street
Los Angeles, CA 90023
213-268-2627
Sidney Cooper

Silverado Foods
512 N John Street
Palestine, TX 75801
800-348-3663
Fax: 918-627-7784
Connie

Steve Smith
P.O. Box 66
Portland, OR 97207

Spa Monopole Corp.
Original Spa Waters
45 E. Putnam Avenue
Suite 121
Greenwich, CT 06830
203-629-2781
Fax: 203-629-2798
William V. Mimnaugh

Edward Spiro
Edward Spiro Photography
340 West 39th Street
New York, NY 10018
212-947-7883
Fax: 212-967-0840

Starbucks Coffee Company
2401 Utah Avenue S, 8th floor
Seattle, WA 98134
206-447-1575
Cheri Libby

Stretch Island
P.O. Box 570
Grapeview, WA 98546
206-275-6050
Frank Kibler

Studio Bolo
4764 55th SW
Seattle, WA 98116
206-933-1157
Fax: 206-933-1191
James Forker, Cindy Winemiller, Carolyn Vibbert, Kit Hutchin

Studio Bustamante
2400 Kettner Boulevard #226
San Diego, CA 92101
619-234-8803
Fax: 619-234-1807
Gerald Bustamante

Sullivan and Brownell, Inc.
Box 450
Randolph, VT 05060
802-728-3300
Fax: 802-728-3309
Thomas Brownell
Kent Gardner

SullivanPerkins
2811 McKinney Avenue
Suite 320, LB#111
Dallas, TX 75204
214-922-9080
Fax: 214-922-0044
Ron Sullivan,Kelly Allen, Paul Brouillette, Art Garcia, John Flaming, Clark Richardson, Linda Helton, Melissa Witt

Supon Design Group
1700 K Street NW, Suite 400
Washington, D.C. 20006
202-822-6540
Fax: 202-822-6541
Supon Phornirunlit, David Carroll, Andrew Dolan, Andrew Berman, Richard Lee Heffner, Apisak "Eddie" Saibua

Sussman Prejza
3960 Ince Boulevard
Culver City, CA 90232
310-836-3939
Jean Campbell

Sutter Home
P.O. Box 248
Saint Helena, CA 95474
707-963-3104
Fax: 707-963-2381

Sweet Adelaide Enterprises
12918 Cerise Avenue
Hawthorne, CA 90250
310-970-7480
Fax: 310-970-9809
Paula Savett

Tait Farm
RR1 Box 329
Centre Hall, PA 16828
814-466-2386
Fax: 814-466-6561
David B. Tait,
Kim Knorr-Tait

Talking Rain
P.O. Box 549
Preston, WA 98050
206-222-4900
Fax: 206-222-4901
Doug MacLean

Tana & Company
511 Sixth Avenue #370
New York, NY 10011
212-633-1910
Fax: 201-655-0219
Tana Kamine

Target Stores
(Archer Farms)
33 South 6th Street
Minneapolis, MN 55402
612-304-6073

Tazo
735 NW 18th Avenue
Portland, OR 97209
503-223-1681
Fax: 503-223-1989
Tal Johnson

Teeccino
1720 Las Canoas
Santa Barbara, CA 93105
805-966-0999
Fax: 805-966-0522
Caroline MacDougall

Thornton and Associates
2115 Northland Drive, Suite A
Austin, TX 78756
512-451-1051
Fax: 512-451-0999
Gary Thornton

Tim Boole Studios
2041 Farrington Street
Dallas, TX 75207
214-742-3223
Fax: 214-748-7625

Toucan Chocolates
P.O. Box 72
Waban, MA 02168
617-964-8696
Michael Goldman

Tudhope Associates Inc.
284 King Street East
Toronto, Ontario
M5A 1K4
Canada
416-366-7100
Fax: 416-366-7711

Twin Valley Popcorn
427 Commercial
Greenleaf, KS 66943
913-747-2251
Fax: 913-747-2424
Ed Henry

Tyler Boley Photography
911 East Pike
Suite 333
Seattle, WA 98122
206-860-7166
Fax: 206-860-0174

Vermilion Design
2595 Canyon Boulevard
Suite 350
Boulder, CO 80302
303-443-6262
Fax: 303-443-0131
Susan Aust

Vermont Milk Producers
RD1 Box 120
Whiting, VT 05778
802-897-2769

Vignelli Associates
475 Tenth Avenue
New York, NY 10018
212-244-1919
Fax: 212-267-4961
Massimo Vignelli

Volcano Island Honey Co.
P.O. Box 1709
Honokaa, HI 96727
808-775-0806
Richard Spiegel

VSA Partners, Inc.
542 South Dearborn Street
Suite 202
Chicago, IL 60605
312-472-6413
Fax: 312-427-3246
Dana Arnett, James Koval, Mary Frank Lempa

Werner Design Works, Inc.
126 N Third Street #400
Minneapolis, MN 55401
612-338-2550
Fax: 612-338-2598
Sharon Werner, Todd Bartz

Werremeyer Creative
15 N Gore
Saint Louis, MO 63119
314-963-0505
Fax: 314-963-0677
Gretchen Floresca, Bob James, Doug Graham

West End Products
4814 Washington Boulevard
Suite 310
Saint Louis, MO 63108
314-361-1138
Fax: 314-361-7178

West Fork Creations
P.O. Box 27
Red Lodge, MT 59068
406-446-3060
Fax: 406-446-3070
Richard G. Poore

Wieden & Kennedy
320 SW Washington Street
Portland, OR 97204
503-414-3781
Todd Waterbury, Peter Wegner, David Cowles, Daniel Cowles,Calef Brown, Charles Burns

Zeus Mediterranean Foods
P.O. Box 13482
Charleston, SC 29422
803-827-1812
Mike Burkhold

Zipatoni
1017 Olive Street
Saint Louis, MO 63101
314-231-2400
Tom Corcoran, Lingta Kung

Zu Design
150 Chestnut Street
Providence, RI
401-272-3288
Fax: 401-273-0633
Taro Masuda

Index

Products